W9-BUM-449

AMERICAN BUILDING

1: The Historical Forces That Shaped It

AMERICAN BUILDING

1: The Historical Forces That Shaped It

BY JAMES MARSTON FITCH

SECOND EDITION REVISED AND ENLARGED
ILLUSTRATED

HOUGHTON MIFFLIN COMPANY BOSTON
THE RIVERSIDE PRESS CAMBRIDGE ·

Second Printing R

Library of Congress Catalog Card Number: 65–10689
Printed in the United States of America

To C. R. and J. D.
who at different levels
made it possible

PREFACE TO THE SECOND EDITION

A field as large and complex as that of American building confronts the historian with a paradox: he cannot hope, in a single work, to be both comprehensive in scope *and* exhaustive in treatment. This was the case in 1947, as I pointed out when the first edition of this work appeared. It is more than ever true nineteen years later. Fortunately, however, the intervening years have seen the coming of age of architectural historiography in the United States. Many trained and competent scholars have entered the field and are now at work on a wide range of specialized problems. A large and rapidly expanding corpus of research on men and movements is available in periodical, monograph and book form. Such a rich literature at once suggests and facilitates general surveys such as mine and has encouraged me to undertake this revised edition.

In the Preface to that earlier edition I wrote the following paragraphs, and these appear to me as pertinent today as they did then:

"Since building is, next to agriculture, America's largest production field, it is obvious that a great many Americans are directly involved in it. To isolate this field in its exact extent, however, is not easy. It is large, complex, and loosely organized; and the interests of the millions in it are varied and contradictory. Thus, to architects and engineers a building is something you design. To a manufacturer of building material, it is a market for your product; to a carpenter, a source of employment; to a mortgage banker, a field for investment; to a landlord, a source of income. Building can be analyzed from the point of view of each of these groups — in fact, it is constantly being so studied. Each analysis always reveals the special interests of the group involved; and though it may sometimes be difficult to realize that they are all talking about the same thing, one fact at least is clear: all of these groups make their living out of the building field. In this sense they have a common identity as producers, and as such their problems differ very little from those of agriculture or transportation, or indeed of society itself.

"But these groups have another and equally important identity and one which they share with all the rest of us — that of being consumers of building. And it is from this point of view, the consumer's, that building may be most fruitfully analyzed. For it is only from such a perspective that building appears as something much more than merely a thing you make drawings of, or sell your products to, or invest your money in, or get your income from. In short, it is only from the consumer's point of view that the social function of building can be fully seen and understood.

"To a greater extent than perhaps any other nation, we Americans have become an 'indoor' people. A large portion of our lives — working, sleeping, playing — is spent in buildings: buildings over whose design and construction we have little or no control; buildings whose physical and economic distribution are only remotely conditioned by our needs; buildings whose effect upon our health and happiness is only obscurely understood. Yet the impact of American buildings upon every aspect and area of American life can scarcely be overemphasized. They are absolutely indispensable tools for controlling our environment, without which modern life would be impossible. We must pay (and dearly) for the use of these buildings, in transactions much like those involving any other commodity; and, to a large extent, they condition our physical and mental well-being.

"Yet in daily life this umbilical relationship between men and buildings goes largely unremarked. It takes a very large building failure, collapse, or fire to rate the same newspaper space as the smallest murder. The news-reel emphasis is always much more upon the unpredictable violence of the cyclone than upon the easily predictable failure of the building; while the woman who falls to her death down a badly designed stairway is merely reported to have died 'as the result of a bad fall.' The chronic shortage of adequate housing calls forth little more comment than bad weather, while the congestion and ugliness of our cities remain, if not largely unnoticed, then at least largely uncorrected.

"Very successful buildings do not fare much better. The more suitable is a building for the use to which it is put, the more unobtrusive is it apt to become to its users, the more subtle the dividends it yields its tenants. It is difficult for the ordinary family to describe the beneficent impact upon its daily routine of a well-designed and well-equipped house — not because the improvement is not real, but because it is hard for the layman to measure or describe it. Well-being, comfort, and health seldom make good headline material.

"Millions are involved in the production of buildings: all of us use them. Yet, despite this fact, few of us are able to tell a good building from a bad one or realize the importance to us of the difference between them. American building today shows immense potentials; it also has great deficiencies. To be able to discriminate between the two is thus a question of first-rate importance to everyone. For here, as elsewhere, an informed public is the prerequisite to closing the gap between what we *could* do and what we *are* doing. It is in the hope of aiding this process that I have written this book."

While a new edition has offered me the opportunity to correct many sins of omission and commission in the first, I have not found it necessary to alter either the point of view or the basic format of the work. Both seem to me as valid now as then. However, to handle adequately new developments both historical and theoretical, I have found it necessary to enlarge the work considerably. This has led to the decision to publish this new edition in two separate volumes of which this, the historical, is the first. The theoretical will follow, I hope, within the year.

James Marston Fitch

Columbia University in the City of New York
January 3, 1966

CONTENTS

AMERICAN BUILDING

1: The Historical Forces That Shaped It

1. 1620–1776
WHAT WE HAD TO BEGIN WITH

By some strange elision of historical and geographic fact, "Colonial" architecture has become a rosy legend of neatly bricked physical comfort, stretched to cover the entire period from the days of the Pilgrim Fathers down to those of *Gone With the Wind* and extending from sea to shining sea. The actual picture is more complex, for the extent of this style was much smaller in time, space, and class. The term, if it means anything, refers to the relatively standardized system of structure, plan, and ornament in use from about 1700 up to the founding of the Republic. It was largely confined to the English colonies along the eastern seaboard. It was preceded in time by the earlier Spanish building in the Southwest; by the early Dutch settlements in New Amsterdam and upper New York; and — most importantly — by almost a century of experimentation with, and adaptation of, late medieval building theories along the eastern seaboard itself. It coexisted in time with the thriving colonies of the French along the Mississippi and with the Spanish in the Southwest. It was followed by a whole series of Classic idioms. And under it everywhere and all the time, like bedrock, the common people continued the while to erect their own buildings, following inherited structural concepts of their own and using materials which were nearest to hand.

However, if it serves no other purpose, the reconstructed Governor's Palace at Williamsburg (Fig. 23) does dramatize the extraordinary accomplishments of that first century of colonization. Together with such buildings as the State House in Philadelphia, or

St. Paul's in New York, it establishes the fact that a building technology qualitatively as high as that of Europe had been rooted in the Wilderness — that skilled artisans, processed materials, and appropriate structural theories were generally available in the urban centers of the eastern seaboard. How remarkable an accomplishment this was is best demonstrated by conditions in the earliest days of the colonies.

One has only to read the fragmentary accounts of the early settlers themselves to realize that they left nothing in Europe and certainly found nothing here which resembled "Colonial" architecture. Theirs was a wretched lot during their first days on these hostile shores. Nothing in their European experience prepared the settlers for the rigors of a New England winter or the heat of a Virginia summer. Everywhere there were extremes of heat and cold, of drought and rain, mosquitoes, flies, gnats — and hostile Indians. They had, according to one chronicler, to

> burrow themselves in the Earth for their first shelter under some hillside, casting the Earth aloft upon timber; they make a smoky fire against the Earth on the highest side; and thus these poor servants of Christ provide shelter for themselves, their wives and little ones, keeping off the short showers from their lodgings but the long rains penetrate through to their great discomfort in the night season.[1]

Again and again, the acute discomfort of these "poor servants of Christ" is mentioned in early documents and the importance of adequate building stressed. Captain John Smith describes a Virginia church which is far different from those handsome buildings in which Williamsburg worshiped a century or so later:

> When I first went to Virginia, I well remember we did hang an awning to three or four trees to shadow us from the sun; our walls were rails of wood, our seats unhewed trees til we cut planks, our Pulpit a bar of wood nailed to two neighboring trees. In foul weather we shifted into an old rotten tent; for we had no better. . . . This was our Church, til we built a homely thing like a barn, set upon Cratchets, [and] covered with rafts, sedge and earth; so was also the walls. The best of our houses [were] of the like curiosity but for the most part far worse workmanship, that could neither well defend [from] wind nor rain.[2]

In other words, in both New England and Virginia, the colonists huddled at first in sod huts and dugouts which were technologically at a lower level than the Indian's bark and bent-wood "long house."

An environment so overwhelmingly hostile to both individual and social life left no alternative but the creation — as rapidly as might be — of a base of operations in which the colonists could lay their plans, husband their strength, and sharpen their tools for the conquest of a continent. They had to build: and build they did, "driven by hunger and haunted by fear," with a savage determination which had not been seen before and would seldom be seen again. The rich, dark continent lay before them: lust and necessity jointly drove these early settlers forward.

Unlike the English and French in Southeast Asia or the Spanish in Mexico, European colonists in North America found neither native populations which they could enslave nor native building tech-

Fig. 1. Houses, Salem, Mass., c. 1630. Modern reconstructions of early Puritan "wigwams" — bent-pole frames, lashed together and covered with bark — and a thatched cottage following European types.

nologies which they could appropriate. The Indian tribes of the Northeast were still nomadic hunters; and while the Roanoke Indians of the Virginia coast, like the Seminoles of Florida, had settled cultures, with agriculture, architecture, and village planning of a relatively high order, little of it proved directly applicable to the needs of the colonists. The Indians of the Pacific Northwest had a quite highly developed wooden architecture, even a rude kind of carpentry: but American contact with them did not occur until the nineteenth century, when aboriginal techniques would have seemed mere obstacles to expropriation and white settlement. In the Southwest, the Spanish overran the cliff dwellings and planned valley cities of the great Pueblo peoples. Although this culture was long past its prime, its mud masonry building technology was of a quite high order, suitable to the climate and not unlike what the Spanish knew at home. Here, and here alone, the European settler was able to put the native population to work.

Thus, of all the many areas of the world colonized by Europeans since the sixteenth century, North America was one of the few which offered no indigenous architecture which could be co-opted and cultivated for the colonist's use. The result was that our architecture begins with the importation of building concepts and techniques developed in other climates and conditions and not especially well suited to those occurring here. Two characteristic tendencies were thereby set in motion: the need for the speedy invention and development of architectural forms suitable for New World conditions; and the necessity for the colonists to do all the work themselves. The result was an architectural tradition of immense plasticity, on the one hand; and a chronic shortage of labor which drove toward rationalization, standardization, and industrialization of the building process, on the other. The U.S. Patent Office presents an impressive record of the results of this peculiarly American condition. No machine was ever too intricate to build, no manufacturing process ever too ambitious, if it promised to reduce the unit cost of labor by increasing its unit productivity. This central economic fact has conditioned our building activity from the very start, giving our architecture its most admirable as well as its least attractive features.

Fig. 2. Indian village near Pamlico, N. C., c. 1586. First English expeditions discovered well-established agricultural communities. Houses were of bent poles, lashed together and covered with bark.

Fig. 3. Seminole Indian village, Florida East Coast, c. 1564. Spanish expeditions found Indians living in well-planned and tidily-kept villages surrounded by wooden palisades for defense.

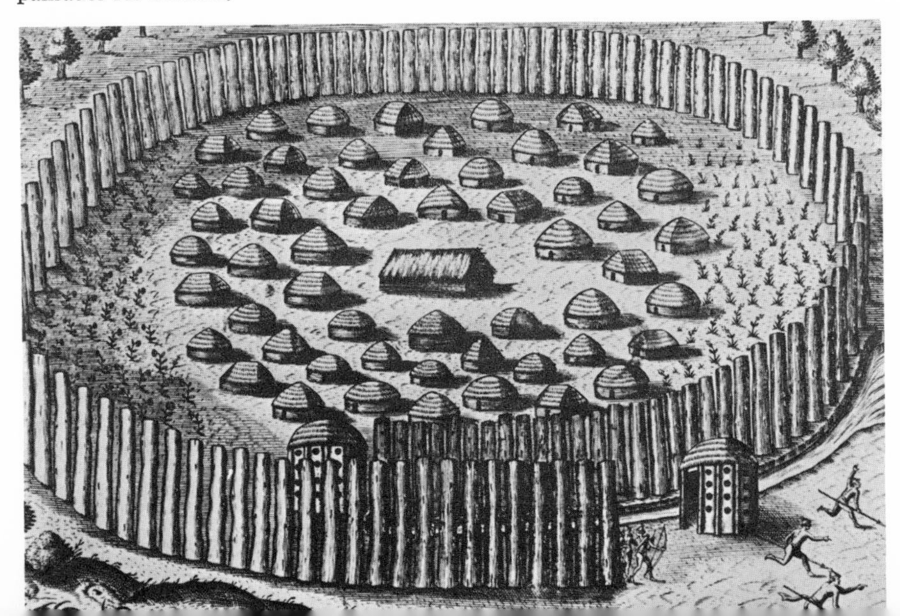

Fig. 4. Museum habitat group showing basswood industry of the Sauk and Fox Indians in Midwest.

Fig. 5. Blackfoot Indian Tipi of Great Plains: demountable shelters were needed by migrant hunters.

Fig. 6. Mesa Verde, Colo., 1100–1300 A.D. Mud-walled, multi-story village for a settled agriculture built under overhanging, south-facing cliff for protection against weather and attack.

Fig. 7. Mingit village, Sitka, Alaska, 1869. A mild climate and abundant "harvests" of spawning fish supported a wealthy neolithic culture. Even without metal saw or ax, Northwest Indians built well using fine woods of primeval coniferous forests.

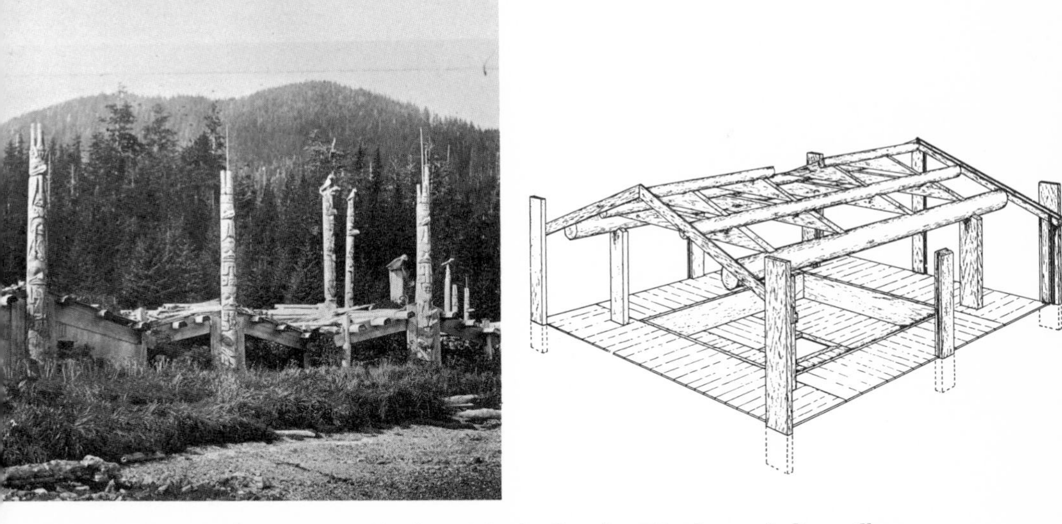

Fig. 8. Tanu Village, Queen Charlotte Island, Canada. Northwest Indian villages, made up of wood-built row houses, were always located on sheltered coves and protected from wind and weather. Elaborate polychromed wood carving on totem poles and house fronts shows relative wealth of the society.

Although practically all the settlers arrived here as members of planned, commercially inspired expeditions, it would appear that the first ships not only forgot to bring carpenters and masons; they also lacked adequate building tools. This omission was partly the fault of the inaccurate and glowing accounts of the New World which the earliest explorers had brought back. (Captain John Smith had rushed home to publish a best seller in which he had described the climate, soil fertility, wild food, and Indians in terms similar to those associated today with the South Seas.)[3] However, the colonists soon discovered their error, and were not long content with sod hut, lean-to, and hillside dugout. Before the first years were out, they were ordering not mere manpower but specific types of skilled building craftsmen — masons, carpenters, sawyers. They were sending detailed lists of the tools required and allowing each immigrant family a set of carpenter's tools for their home in the New World. Of raw materials they had, of course, more than enough. Their early letters home waxed ecstatic over the vast resources in wood, stone, marble, slate, sand, and clay. However, access to these raw materials was one thing. Converting them into processed building materials was quite another.

STRUCTURAL INNOVATION BEGINS

The level of building in any given time and place stands in direct relation to the specific technological level of society — that is, to the types of skills, tools, and materials available to the building field. This was as true for the early settlers as it is for us; and the rapidity with which they established an adequate material base for their building is one of the most remarkable of their accomplishments. A brick is a simple thing only to a society which has an adequate supply of skilled masons, brickmakers, and brickkilns; to the early settlers it was a precious object. A sawn two-by-four timber is insignificant only to people who do not know what the world was like before the advent of the power-driven saw; and a building of these two products — brick and sawn lumber — is a social achievement whose true significance can only be grasped by people who have endured a smoke-filled dugout through a Massachusetts winter.

By and large, all of the colonists on the Atlantic seaboard were familiar with one or more of the then standard structural systems — wood framing, brick and stone masonry. The great all-masonry construction of the Gothic periods had never been a genuinely popular building technology in the sense that peasants as well as prelates could use it in their domestic buildings. The Gothic style did not appear in America until the early nineteenth century — and by that time it belonged much more to literature than to building technique. Wood, however, was essential to all these systems, since no other suitable tensile material for framing floors and roof was available — then or for the next two hundred years. Although uncut stone was widely used in some parts of the colonies, it was never to become a real competitor of brick on any but a local scale. The brick's small size, standardized production, and general availability made it the pre-eminent masonry material.

But wood was to become America's principal building material, far outstripping all others in the scope and versatility of its use. Though the English unquestionably brought with them the rudiments of our modern wood-framing systems, they were crude and inexact — straight from a feudal England still largely built of wood. (The London which was leveled in the Great Fire of 1666 was a wooden city.) Moreover, English wood frames soon proved to be unsuited to our climate, whose violent and abrupt changes caused rapid expansion and contraction. This led to cracks which reduced the efficiency of the wall. Under the impact of their new surroundings, the Pilgrims modified their structural systems to meet their new thermal environment. At first they merely covered the medieval brick nogging with which the English had filled in the wood skeleton with two relatively impervious skins — the clapboard exterior and the plaster interior. Ultimately, the brick disappeared altogether. Naturally, the new system was not perfected overnight: it was rather a process of small changes here and there, which resulted in new and characteristically American systems of wood construction.

Since wood was at once the most abundant and easily worked material in the new country, and since the settlers were thoroughly familiar with it, it is not surprising that they at once adopted it as their favorite material. What is surprising, however, is the *way* they used it. They apparently dismissed the Indian's bark-covered sapling skeleton as unsuitable from the first. They early discarded the "stockade" — walls of sharpened logs driven into the ground and interstices daubed with mud — as impracticable. And they never employed the log cabin for the simple reason they had never heard of it. (This latter system — simplest, quickest, and relatively most comfortable of all — was the invention of the Swedes, who brought it with them to Delaware in 1638. Knowledge of it did not spread much beyond there until the eighteenth century.) Instead, the New Englanders turned directly to the wood-framed, clapboard-sheathed structure which is today called "saltbox" (Figs. 12, 13).

In many ways this was a remarkable development. It involved attacking the building problem at the point of greatest resistance. For while this system provided the most efficient structures, it also implied the most advanced equipment — specifically, power-driven saws. Both the English and the Dutch were accustomed to pit-sawing — long, straight saws operated by two men, one above, one below. But neither were familiar with sawmills, which, though known in England, were outlawed for fear of technological unemployment among the sawyers. So at first the early settlers resorted to pit-sawing. However, they soon found that hand labor was far too

Fig. 9. "The Oaks," West Bromwich, England, c. 1550. The heavy, hand-hewn oak frame of English medieval house was in-filled with brick or wattle-and-daub. Moderate climate kept movement of frame to minimum.

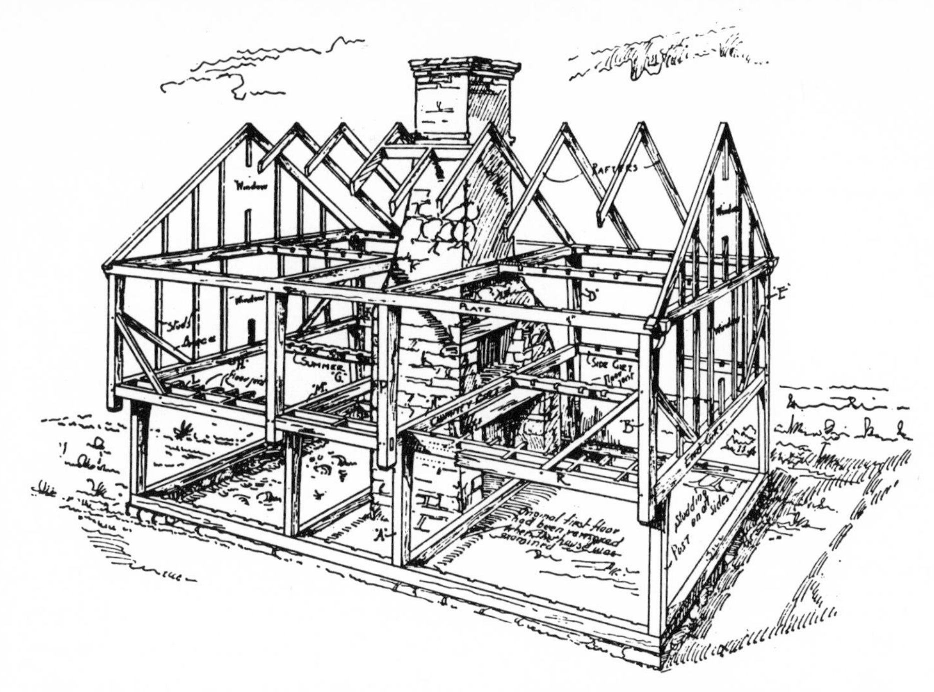

Fig. 10. Gleason House, Farmington, Conn., c. 1650–1660.. This structural analysis by Isham and Brown shows how New England houses followed English precedent. Harsh climate soon showed vulnerability of system.

Fig. 11. Harvard College, Cambridge, Mass., 1638–1642. This modern reconstruction of first building shows late-medieval profile and chimneys; but wooden skeleton has been sheathed in weatherproof clapboards.

Fig. 12. Wildman House, Guilford, Conn., c. 1668. Settlers soon developed basic "saltbox" house form. Compact plan, low ceilings, limited fenestration and huge central chimneys were standard features.

Fig. 13. Baldwin House, Branford, Conn., c. 1645. The "saltbox" saw an evolutionary development (below and right), in which a series of modifications made it steadily larger, more comfortable and more commodious.

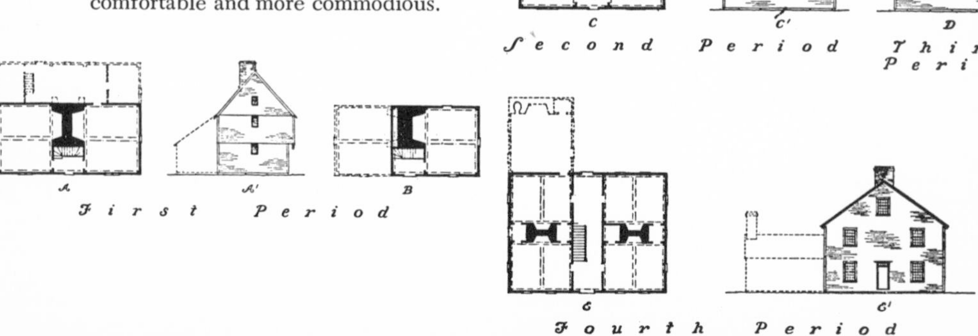

slow and costly for their need and — in early but typical American fashion — mechanized the process. Thus, a short thirteen years after the Pilgrims landed, they were establishing their first power saw at the Falls of Piscataqua on the line between Maine and New Hampshire. It was a cumbersome apparatus — a huge swishing up-and-down adaptation of the pit saw — and it must have been the source of much speculation among the villagers. There are unauthenticated reports of even earlier power mills — Maine, 1631, and New York, 1633. In any event, before there were any mills in England, the colonists had mechanized the sawing process.

In the succeeding centuries, a giant lumber industry would arise from these small beginnings, spurred on especially by the appearance of the steam-powered sawmill at the end of the eighteenth century. However, the development of that most American of all wooden structures, the balloon-framed house, awaited the appearance of both standardized sawn lumber *and* cheap nails. Not until the 1830's did the machine-made nail appear: until then, the wooden skeleton had to be assembled with mortise-and-tenon and pegs. These connections required bulky framing members and they were often hand hewn. But, with the appearance of cheap nails, handworked jointing disappeared very rapidly. The entire skeleton was immeasurably lightened and site labor greatly reduced. By 1840, the typical frame house was built completely of milled lumber in standard sizes.[4]

Much the same thing happened in brick manufacture. Possibly as early as 1622, and certainly no later than 1630, bricks were being commercially produced at Powder Horn Hill near Chelsea, Massachusetts. In 1628 the canny Rensselaers were producing the favorite Dutch yellow brick on their estates near Albany, New York, and selling them to all comers for fifteen florins a thousand. Even earlier, in 1611, Sir Thomas Dale had landed in Jamestown and immediately started "many essential improvements" which included a brickyard and a smith's forge. Later the same year the colonial secretary reported that Henrico (near the present site of Richmond) boasted "three streets of well-framed houses, a handsome church, and the foundations for a more stately one laid in bricks." Although not founded until much later, New Orleans early provided for the es-

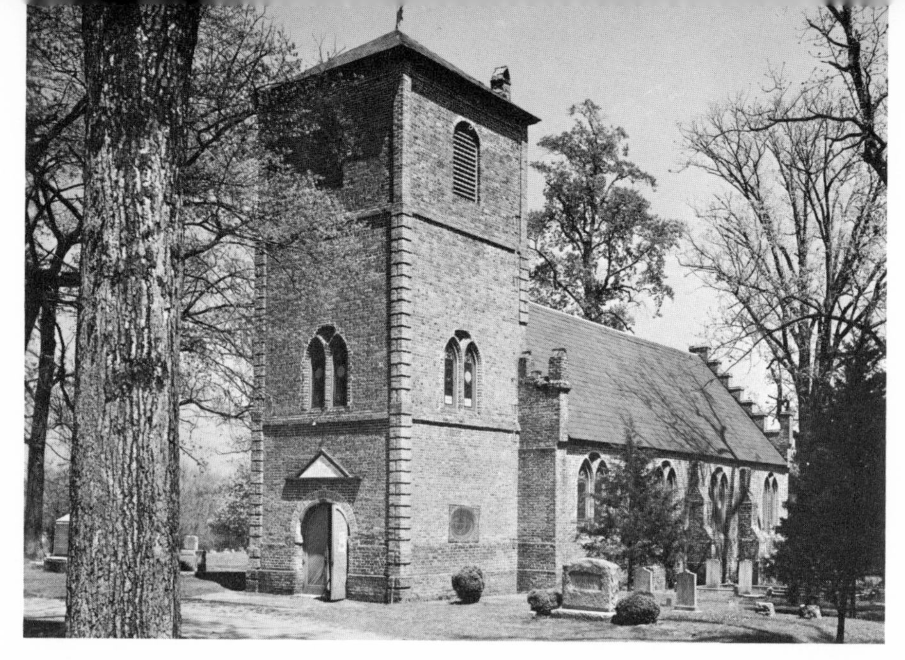

Fig. 14. St. Luke's Church, near Smithfield, Va., c. 1631. This modern restoration probably gives a fairly accurate picture of the late-medieval aspect of early vernacular building activity.

Fig. 15. The Parsonage, Albany, N. Y., c. 1657. This stepped-gable brick town house closely followed the urban Dutch prototype. Accustomed to brick in Holland, the Dutch used it widely in New York.

tablishment of brickkilns which were to be worked by convict labor from among the salt smugglers.

It is thus clear that the establishment of a material basis for the building field was the first order of business for all the colonies. Since the problem was too urgent to be left to individual initiative, early records are replete with decrees, subsidies, and interventions promoting such ventures. Even then, in a period when every man was of necessity somewhat skilled in many crafts, building labor was at a premium. Some hint of the importance of building workers in colonial economy is revealed by a choice bit of Massachusetts anti-labor legislation dated August 23, 1630:

> At the First Court of Assistants holden at Charlton [Charlestown], it was ordered that carpenters, joyners, bricklayers, sawyers, and thatchers shall not take above 2 s/ a day, nor any man shall give more, under pains of 10 s/ to taker and giver.[5]

Despite the policy of the British Crown of discouraging colonial industry, and the consequent necessity for importing glass, hardware, nails, etc., the colonists had established their own brick and lime kilns, sawmills, small foundries, and glass factories long before they gained independence.

PUT ANOTHER LOG ON THE FIRE

It is still not generally recognized that if the Pilgrims had landed on a nearby planet instead of the New England coast, they could scarcely have made a more abrupt switch in thermal environments. In Plymouth, England, they left a moderate climate, with a very stable temperature, without extremely cold winters or very hot summers; snows and droughts were rare; tornadoes and cyclones unknown. In Plymouth, Massachusetts, they found a thermal environment whose annual cycle was far more severe, with a temperature spread from July to December more than twice as great as in Plymouth, England. They found also heavy snowfalls, long freezes, enormous gales.

The heat of the New England summers, though doubtless uncomfortable to the settlers, was not actually dangerous to their health. They could always sit down in the shade. But the severity of the

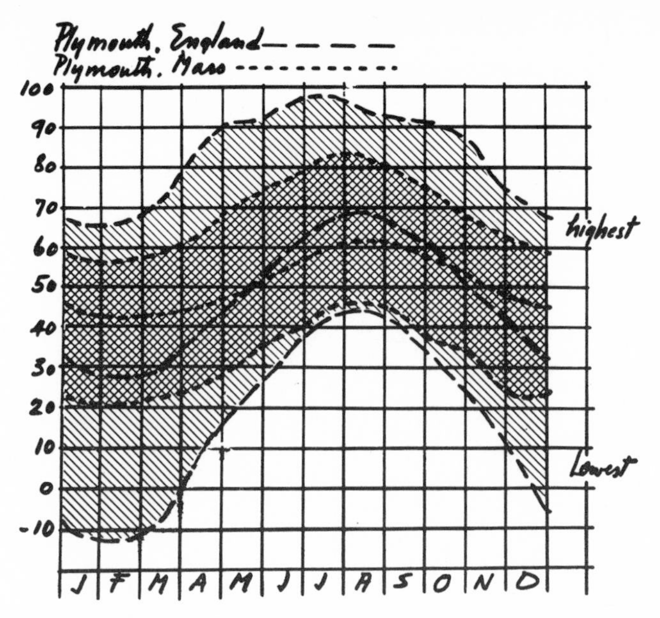

Fig. 16. When compared, the annual temperature regimes of Plymouth, England, and Plymouth, Mass., reveal great divergencies. The New World has much colder winters, much hotter summers and much greater diurnal temperature differentials. Hence destructive movements in building frames are more severe.

winters, with the consequent need for adequate heating sources, was a far different matter. The urgency of this problem is attested by their early buildings, the most characteristic feature of which was always the huge chimney with its fireplaces (Fig. 17). Since these served also for cooking and baking, they literally became the heart of the house. Within the limits given, these huge chimneys were cannily designed. The fireplaces themselves were none too efficient; but the mass of the chimney — located in the center of the house and occupying a disproportionate amount of floor space — absorbed the heat of the flue gases and radiated it into the house. They were far from ideal as heating media and — romantic legend notwithstanding — cooking on them was backbreaking drudgery. So it is not surprising to find the fireplace beginning to be supplanted by the stove, as soon as the rising technological level of colonial society made that possible.

The Pilgrims had already modified their structural systems to make their buildings weathertight under the new conditions. They also

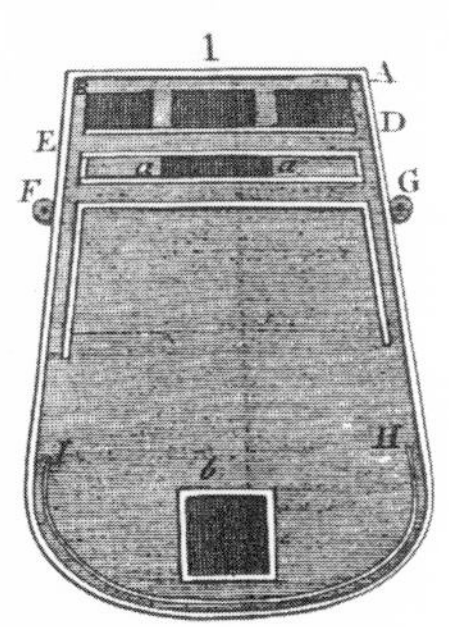

Fig. 17. Kitchen from Seth Story house, Essex, Mass., c. 1684. This installation at Winterthur Museum, near Wilmington, Del., gives some picture of late-seventeenth-century cooking technology. Laborious and dangerous work, it was not to be substantiallly eased until appearance of iron cook stove.

Fig. 18. Prefabricated cast-iron heating stove designed by Benjamin Franklin, c. 1735. Note the simplicity of standardized elements; they could be shipped flat and quickly assembled by buyer.

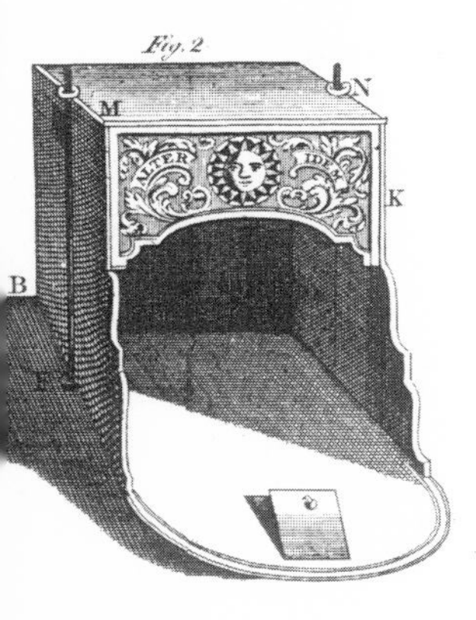

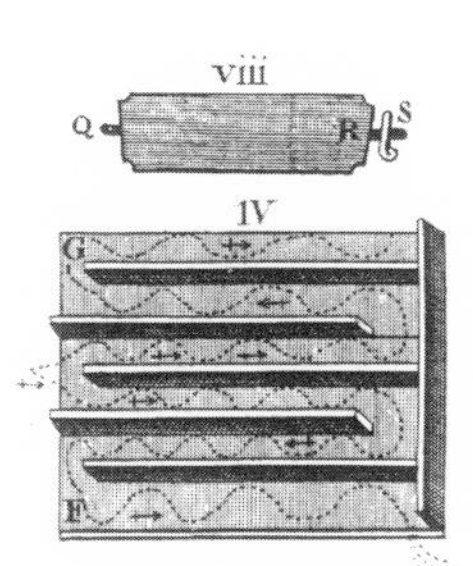

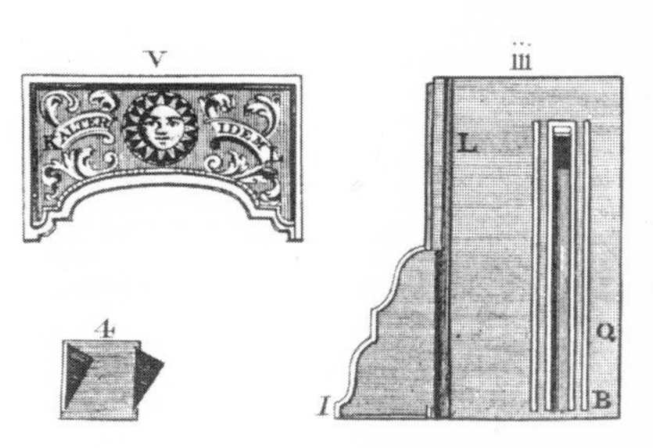

sought to make their heating more efficient by small rooms with low ceilings and small windows. (This last was an economy in another sense, for glass was dear and hard to get, and both the British Crown and local governments had quickly slapped a tax on it.) For such features as these, early American building deserves no special credit. They were common practice abroad. But when Benjamin Franklin's stove appeared in Philadelphia in 1744 — a scant sixty years after William Penn had cleared away the forest — a decisive step had been taken. This stove, like an even earlier but commercially unsuccessful one by Christopher Sower of Germantown, differed in many important respects from its cumbersome European ancestors. To begin with, it was prefabricated. Using the most modern material of his time — cast iron — Mr. Franklin was able to turn out on a mass-production basis a stove which could be readily assembled from a minimum number of standardized parts. It was lightweight, self-contained, efficient, and relatively cheap. One has only to compare it with the ponderous masonry units of Europe to understand how truly revolutionary were its implications.

In its earliest form, the Franklin was half fireplace, half stove. It had no doors and its shape, grate, and raised hearth showed clearly its masonry ancestry. Its principal significance lay in the fact that it was detached from the chimney. It could be installed anywhere because it was lightweight and self-contained; and with the addition of a door, it became a stove. From this early innovation of the indefatigable Franklin, all modern stoves and furnaces and most air conditioners are lineally descended. For heating purposes, these stoves marked such an advance over the fireplace, both in terms of efficiency of conversion and efficiency of heat distribution, that they soon outstripped it. For cooking purposes, of course, their advantages were even more pronounced. In fact, it would be interesting (though not germane to this study) to trace the decisive effect of the "range" on American cookery. The slow heat of the masonry oven had led naturally to the breads of Europe, just as the size, cost, and technique of the oven necessitated the communal baking of bread. Contrariwise, it is apparent that the stove's quick intense heat has profoundly affected our dietary habits and made possible the invention of such American breads as the biscuit, corn pone, flapjack, and waffle.

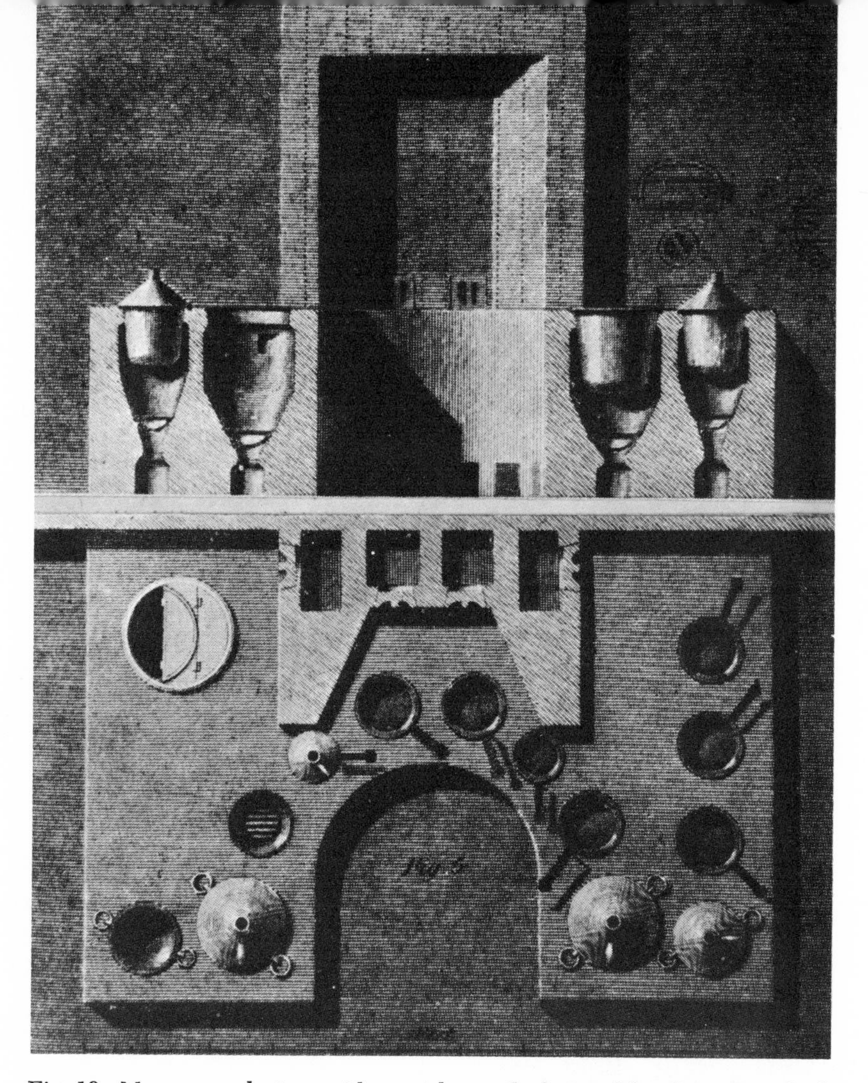

Fig. 19. Masonry cook stove with special utensils designed by Count Rumford, c. 1790. An early investigator of problems of hygiene and nutrition, Rumford's stoves provided chef with vessels in a variety of sizes and temperatures, arranged in a convenient, waist-high semicircle.

The problem of ventilation, always closely associated with that of heating, also did not escape the attention of early Americans. It was naturally more urgent in the intense heat of the South, and as the Carolinas and Virginia developed there was a steady increase in ceiling heights and window sizes. Here, too, we see the beginnings of another American invention — one might almost say institution: the porch. This grew steadily in size and importance as one moved south until it became the dominant aspect of upper-class residential structures. Imagine Mount Vernon without its portico! In New Orleans, Spanish familiarity with balconies and arcades led to bril-

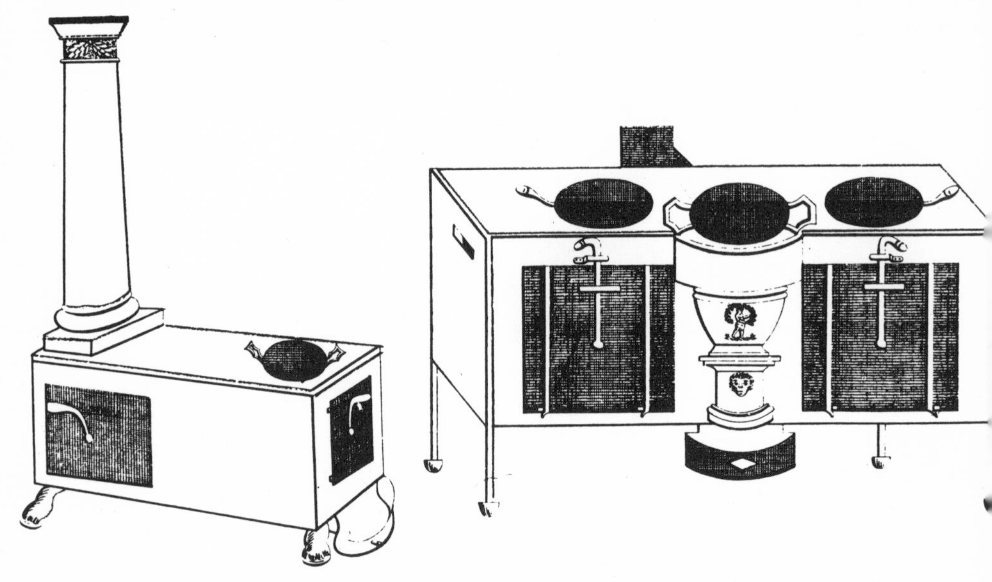

Fig. 20. Cooking and heating stoves, c. 1820. These prefabricated metal stoves, manufactured in Holland by J. F. Dudy, were imported in quantity into this country. Despite Empire detail, they anticipate the modern cooking range in general design.

Fig. 21. Laundry and cook stoves manufactured in Connecticut. Although these date from the 1850's, the basic types had been perfected and manufactured everywhere by the 1840's. All early stoves burned wood; coal-burners appeared after the Civil War.

liant adaptations in masonry and iron (Fig. 81). These two streams were ultimately to meet in the reactionary ostentation of antebellum plantation houses, with their slave-powered, peacock-feathered punkahs, twenty-foot ceilings, and continuous many-storied galleries (Fig. 83).

The more subtle problems of fresh air in heated rooms, humidity control, and elimination of drafts had not yet risen to dog the colonists. Yet there are some prophetic signs of future advance in atmospheric and thermal control. Thus we find Benjamin Franklin taking time off from the press of diplomatic work to describe to the physician to the Emperor at Vienna a ventilation device he had noticed in France:

> Take out an upper pane of glass in one of your sashes, set it in a tin frame giving it two springing angular sides [that is, like the letter slot in a modern post box] and then replace it with hinges below. . . . By drawing this pane in more or less you may admit what air you find necessary. Its position will naturally throw that air up and along the ceiling.[6]

That so simple a device should have been the subject of correspondence between two such eminent scientists is an index of the general level which obtained in the field of heating and ventilating.

Those who incline nostalgically back to the days when gilt and crystal chandeliers shed soft light (and dripped hot tallow) on the heads of colonial dancers confuse the image with the reality. The candle which was kind to the faded beauty or the jaded swain was the same which set many a house and a hoopskirt afire. Besides being smoky and unsafe, candles were expensive; there were probably not a hundred houses in the thirteen colonies which could afford the candles necessary to illuminate properly a ballroom for an entire evening. The vast majority of early Americans either made their own candles as they did their soap or else they studied, sewed or danced in the light of log fires or flaming pine knots. For the lamp chimney had not yet been invented and the whaling fleet had not yet appeared to furnish cheap oil. Those candlelit drawing rooms which ornament our Christmas cards and calendars thus give little hint of the reality of colonial illumination — of dark, muddy streets and dangerous coasts without lighthouses, of workshops "dark in the blaze of noon," and cabins without glazed sash, so that even on the brightest winter day the wooden shutters were kept closed.

Personal hygiene was still fogged in medieval ignorance and superstition. When teeth were brushed at all, it was with table salt and a chewed twig. Soap was already a staple item in every household, but it was apparently more used on the family laundry than on the family itself. Bathing the whole body was still considered dangerous, even though Benjamin Franklin had already introduced a shoe-shaped sitz bath and was cautiously advertising the benefits of bathing.

Putrescence dogged the housewife in her efforts to feed the family. The technology of food preservation by canning was still three-quarters of a century away (the Mason jar was not perfected until 1858). Refrigeration of food by icing had not yet been discovered: thus milk could only be preserved over any period in the form of cheese; and meats could be held over only by salting or smoking. Jams and jellies were a known way of preserving fruit but the high cost of West Indian sugar made them a luxury only the urban rich could contemplate. Honey or maple sugar furnished what sweetening most farm families would have had from one year to the next. Vegetables, except for roots such as turnips or potatoes, would have been completely seasonal.

In such a context, sanitation concepts were necessarily primitive. Table scraps were fed to the family pigs or — in the growing towns — dumped in the streets for wandering free-lance pigs to root in. Open-pit privies were universal in both country and town. In New Orleans slop jars were emptied into open storm sewers and drainage ditches. In other cities, human excrement was sometimes collected by mule-drawn carts and hauled in barrels to the open country. (Though, fortunately for colonial health, there seems to have been no use of it as fertilizer.) Sewage-disposal systems were unknown and public water systems were still to appear. Dug wells were the source of most urban drinking water, though farmers built their houses near good springs whenever possible. Manually operated pumps were appearing in the towns by the time of the Revolution, though the bucket-and-windlass was the principal means of lifting water to the surface. Since hard water was so prevalent, there was a fairly wide use of rainwater storage systems — masonry cisterns to which all the household gutters were connected. All in all, with the whole discovery of the connection between polluted water and pestilence still a century or more away, it is remarkable that colonial health was as good as it was. The principal reason was probably the preponderantly rural character of the nation and the small size of the towns.

By the time of Washington's inauguration, the building of the new republic had consolidated itself at about the same technological level as that of Europe. It employed much the same structural theories and building materials, and was by this time largely based on its own industries. It had approximately the same percentage and types of skilled artisans and amateur gentleman architects — both of whom studied the flood of architectural and archeological handbooks which were a phenomenon of the times. If American building projects were fewer and smaller than those of Europe, this reflected more the relative immaturity of colonial society than any absolute deficiency in its means of building.

If the building of the period was abreast of Europe in most respects, it already showed certain prophetic indications of even more rapid advance in others. This was particularly the case in planning. The structure of a given society is clearly revealed in the sum total of its building plans; and the more advanced the society, the more varied and complex are its building types. Long before the Revolution, the trend toward the creation of new building types — that is, new plans — was apparent. Thus, as early as 1735, the plan of the Pennsylvania State House (Independence Hall) showed the qualitative changes taking place in government. Although clearly based on English palace types, its plan reflects the need for large democratic assembly rooms instead of small, kingly audience chambers. The new buildings for the Philadelphia City Almshouse and Hospital in 1760 likewise differed from European prototypes in their expanded facilities for separate attention to aged, poor, insane, and ill.

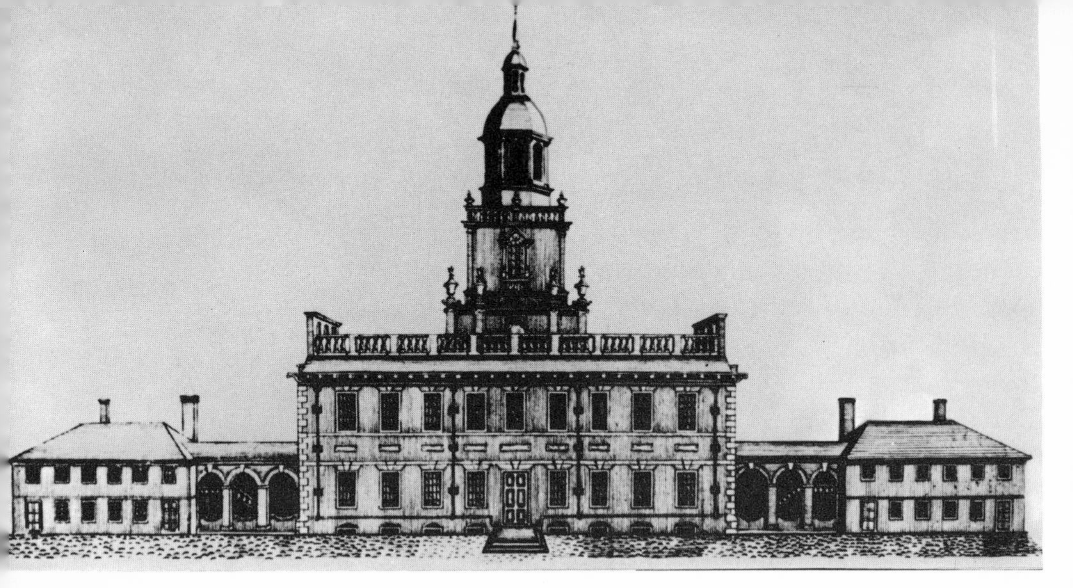

Fig. 22. The State House, Philadelphia, Pa., 1732–35. Edmund Woolley, chief carpenter, probable designer.

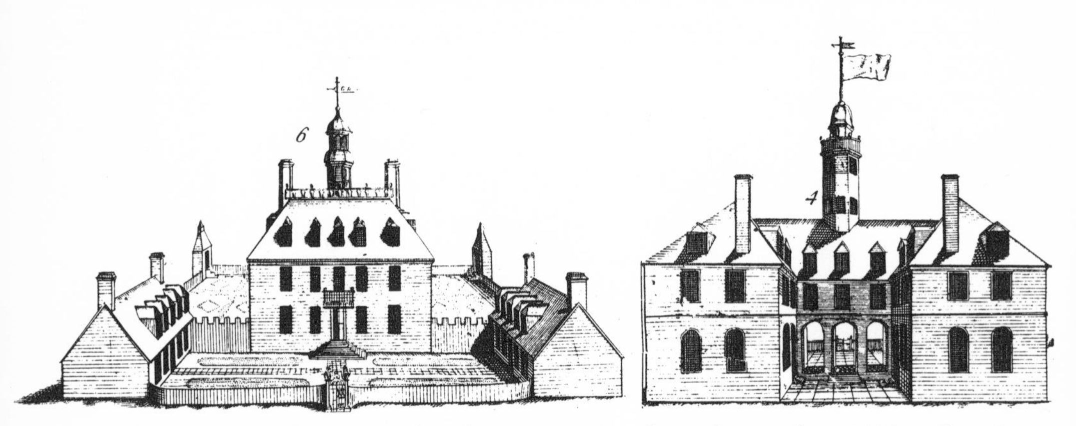

Fig. 23. The Governor's Palace, 1723–47, and Royal Capitol, c. 1745, Williamsburg, Va. These famous drawings from the Bodleian Library in Cambridge, England, were the basis for the modern restoration.

Fig. 24. Charleston, S. C., seen from the harbor c. 1730. The city was one of the largest colonial capitals, exporting rice, indigo and naval stores and importing goods for the hinterland.

American architectural ingenuity, in 1789, would have been best exemplified by what it had accomplished in the free-standing, single-family house. In structure, plan, and volumetric character it had perfected four basic types, each of which was admirably adapted to the climatic, social and economic conditions of its region. These were:

1. The compact, centralized, close-clipped house of New England with its small windows, small rooms, and low ceilings adapted to long, hard winters and relatively cool summers. It could be built in either wood or masonry and readily modified for either farm or urban use (Figs. 10–14).
2. The much more extended plantation house of the Middle Seaboard colonies with its basically Palladian plan. Its hot weather features — high ceilings, larger windows, porches, and porticoes — got progressively more pronounced as it moved south (Fig. 26).
3. The stilted, airy pavilion of the Louisiana French, with its parasol roof and perimetral galleries and balconies to shade the house from sun and rain and its great emphasis on effective ventilation — e.g., floor-to-ceiling windows, high ceilings, central halls. This type was equally urban and rural and equally adapted to wood or brick masonry construction (Figs. 27, 47).
4. The inward-turning, patio-centered, mud masonry hacienda of the Southwest. With thick walls and few windows toward the outer world and larger openings toward the shaded patio, this type was ideally suited to the semi-desert climates of the Southwest (Figs. 29, 30).

Each of these houses was a masterpiece of preindustrial folk knowledge in which locally available materials and techniques were put to sagacious use to modify the environmental extremes of the region. And their viability was proved in the succeeding century by the ease and grace with which they accepted the stylistic nuances of the Federal, Empire, and Greek Revival modes.

In plan, the buildings of colonial America necessarily reflected new conditions; but they also showed the imagination and freedom with which the colonists attacked new problems. This ability to evolve new plans and new building types was the expression on an architectural plane of their fertile ability to evolve new social and economic institutions. The building type which was eventually to dominate the American scene — which was in fact to become the fountainhead of practically every subsequent advance in American building — was still missing: the factory. For this the colonists could scarcely be blamed. Such manufacturing as the British Crown permitted had not grown much beyond the scale of small localized handicraft operations. These imposed no particular strain upon conventional structural theories or plan types. Aside from water-powered flour and lumber mills, and wharfside developments in the big port cities, there was little hint of what was to come.

Yet even in this field, the colonies did not lag too far behind England and Europe. Although the development of power-driven manufacturing was well advanced abroad, it had scarcely begun to reflect itself in buildings especially designed for that purpose. It was only as the eighteenth century drew to a close that the factory began to appear as a recognizable type: and here, from the start, America was destined to excel.

Fig. 25. Parlor, Lee House, Marblehead, Mass., 1768. Wealthy New England mercantile families lived in much the same comfort as their English contemporaries, importing many luxury items.

Fig. 26. "Westover," Charles City County, Va., c. 1737. This typical James River plantation house, built by William Byrd, represents a fairly literal transplant of English country-house design.

Fig. 27. "Madam John's Legacy," New Orleans, La., c. 1750. An urban version of the single-storied, hip-roofed cottage used by the French throughout the Mississippi Valley, but modified for Gulf climate.

Fig. 28. First Court House, Cahokia, Ill., c. 1750. Modern reconstruction.

Fig. 29. Side elevation, San Xavier del Bac, Tuscon, Ariz., 1797.

Fig. 30. Governor's Palace (restored), Santa Fe, N.M., 1609 (?).

THE CLASSIC SPIRAL

When, after days of discussion made no easier by the July heat, the representatives of the thirteen colonies put their final approval on the Declaration which Jefferson had drafted for them, they adjourned to the streets of a new capital which bore a strong resemblance to the London from which they had just seceded. The very hall in which they had met might well have come from the drawing boards of Christopher Wren. The streets and little squares through which they walked might have been laid out, the houses at which these gentlemen were guests might well have been designed by any of the famous Georgian architects. The shops they passed and all the little houses they did not notice could have been the work of any anonymous British craftsman. How did the idiom reappear upon the gaunt New England coast, follow the colonists as they spread south and west, ultimately blanket the nation? The key to this paradox lay in England.

Consistently, since its rediscovery by sixteenth-century England, the Classic world had dominated the English mind. With no serious interruption, it would continue to be a dominant influence for half a century more. The basis of its appeal is apparent: conceptually, the Classic world in all its aspects was symmetrical, rational, concrete. Its large scale and fascinatingly minute detail were equally well lighted by its spirit of free inquiry. For the young bourgeoisie, fighting free of the bemused paralysis and planless growth, the bottomless mysticism into which Europe had sunk, the Classic thus appeared as the ideal instrument for the reconstruction of man's consciousness. The Classic form became a unit of measurement, the standard by which they judged and the mold into which they compressed the most varied expressions of their culture: language, legal system, city plan; façade and poem; sermon, gravestone, and essay.

But the eager Elizabethans could not approach this Classic world directly. Separation in time and space, inadequate scholarship, and limited means of communication prevented a face-to-face meeting. In their lustrous naiveté, they saw it as through a long and twisting hall of mirrors — its outlines blurred, its features magically distorted, its imagery mixed and idealized. Their first perspective was through the magnificent lens of the Italian Renaissance. Shakespeare, like a social seismograph, registered the first repercussions. Small-scale Marco Polos sailed back to England with wondrous tales, books, drawings. The English ruling classes, fascinated by the world they saw, were avid for more information. Foreign sources soon proving inadequate, English scholars began their own translations, explorations, and measurements.[7] Soon the time would come when a knowledge of Roman architecture, art and literature would be an essential equipment for every cultured Englishman.

Generation after generation of Englishmen were engaged in this cultural transfusion. Inigo Jones, beginning as a "picture-maker" to the Earl of Rutland, returned from Rome as an architect in 1614. Sir Henry Wotton first paraphrased Vitruvius — that rediscovered Augustan who became the first great vector of the Renaissance idiom — in 1624. Christopher Wren got no closer to Italy than Paris; but there, in 1666, just a few months before the Great Fire, he met the paladin of Baroque architects, Bernini — an "old reserved Italian" who only allowed Wren a glimpse of his plans for the Louvre.

Eighteenth-century England showed an insatiable appetite for Classic antiquity. Taking time off from a fashionable practice, Robert Adam spent the years 1754–57 in Italy, first studying Roman monuments and Pompeiian ruins; then, with two draftsmen in suite, making his famous drawings of the Palace of Diocletian at Spalato. The works of the German archeologist, Johann Joachim Winckelmann, did much to systematize eighteenth-century knowledge of Roman culture. His monumental studies of 1762 and 1764 afforded the first real information on the current excavations at Pompeii and Herculaneum. Winckelmann's work also directed attention to a yet more distant Classic landscape: the Greek. Based upon inadequate Roman sources, many of the ideas advanced in his *History of Ancient Art* subsequently proved to be erroneous. Nevertheless, his efforts brought the Grecian vista into clearer focus. Another fascinating world was opened to the British mind, the exploration of which would in many respects duplicate that of ancient Rome.

By the close of the eighteenth century, no grand tour was complete without the Acropolis, just as fifty years earlier none had been complete without a moonlight visit to the Forum. Greek was as essential

Fig. 31. Etruscan art as visualized by Johann Wincklemann in his epoch-making study of classic antiquity.

Fig. 32. The Parthenon, as drawn by Stuart and Revett in one of the earliest first-hand reports. Based on actual measurements, the artists aimed at documentary accuracy in their drawings.

as Latin had been earlier. Two Englishmen, James Stuart and Nicholas Revett, spent five years there, making measured drawings of many Greek monuments. When they were published in the handsome book, *Antiquities of Athens,* they generated still another wave of Classic enthusiasm in the English-speaking world. Further to strengthen the sense of immediacy in this curious relationship, Lord Elgin went to Greece. From the Acropolis, under circumstances none too savory, he brought the marble figures from the pediment of the Parthenon back to London in 1816. And Byron, when he died there in 1824 with the Greek armies of liberation, brought the English interest in Greece to a climax. It might be said that, with Byron's death, the journey was complete for the English; the spiral had been traced out, the images disentangled, the Classic mirage explored.

This process of piecemeal rediscovery of the past was reflected with great accuracy in the architecture of England. Indeed, architecture had been from the first one of the principal media of Classical ideology. Already in 1550 it would have been possible to see crude Classic ornament applied to essentially medieval structures. Here anonymous designers, moved by echoes no more precise than those which stirred Shakespeare, began to use the idiom: mask and acanthus, cornice and frieze (the style known today as Tudor). The first great architect to show by his work a firsthand access to reasonably accurate sources was Inigo Jones. In his buildings we find the whole Classic system — column, entablature, and pediment. Nevertheless, Jones saw Rome through the Renaissance lens, as did Sir Christopher Wren, who appeared like a phoenix from the great London fire of 1666. In his plans for the reconstruction of the city and in his design for St. Paul's Cathedral, Wren displayed that grandness of conception and dexterity of detail which marked the Classic at its best. Yet he, too, was a man of the Renaissance, confusing the Rome of Michelangelo with that of the Caesars.

Wren was followed by able men: John Vanbrugh, William Kent, James Gibbs, the brothers Adam. They were facile and well educated, completely familiar with the sources then available. However, these sources were, by modern standards, disconnected and incomplete. They made no clear distinctions between the Rome of

Fig. 33. Entrance porch, "Waterstone," Dorsetshire, England. Architect unknown.

Fig. 34. "Wollaton," Nottinghamshire, England, c. 1595. Robert Smythson, architect.

the Empire and that of the Republic; between metropolitan and provincial Rome; or between archaic and Periclean Greece. This lack of archeological precision was characteristic and probably would not have much disturbed the Georgian architects, even had they known it. For designers such as Robert Adam and his collaborator, the decorative artist Angelica Kauffmann, were more ideologues than scholars. The whole Greco-Roman epoch was a dazzling treasure-trove from which they could borrow at will. Under their influence, the vernacular lost in vigor what it gained in skill. This was in strict accordance with facts of their society. For the Classic concept in the hands of the Georgians had become an instrument of imperialist expansion. Even Byron's death in Greece had two aspects. It had freed the Greeks from Turkish oppression; it had also secured the Mediterranean for British trade with India and the East (Figs. 37, 58–59).

The Georgian idiom flowered and withered, and still the Classicists had not encompassed the whole of the Classic world. Indeed the political decisions which the dissident Americans took in Independence Hall on that July day were destined to produce the most pregnant phase of the entire Classic spiral. For they led to the so-called Classic Revival, the revival within a revival, the white-hot return to simon-pure original sources. This movement was even more explicitly revolutionary than its predecessors had been, and its energy was to be expended much more in America and France than in England. Jacques David and the official artists of the French Revolution were to discover, with stunning timeliness, a very specific and politically concrete Rome, the Rome of the Republic. Jefferson would mirror this enthusiasm in this country and Latrobe would shortly thereafter discover a specifically democratic Greece. And the English intellectuals, alarmed at the literalness with which France and America were pursuing their revolutions, would begin to lose their Classic ardor. Keats would die, Wordsworth would cool, Byron would be submerged in the rising Gothic tide of Scott and Pugin and Ruskin (Figs. 87–90).

Fine as all these stylistic shadings might sometimes appear, they were nevertheless important indices to deeper changes in European

Fig. 35. St. Paul's Cathedral, London, 1675–1710. Sir Christopher Wren, architect. This is the so-called "Great Model," a rejected design (Wren's third) dated 1673.

Fig. 36. Banqueting Hall, Whitehall Palace, London, 1622. Inigo Jones, architect. The Hall was the only portion completed of a much more grandiose complex along Renaissance lines.

and American society. Those Classic perspectives which had enthralled so many successive generations were by no means identical, similar though they may have been in general form. Politically, the movement had reconstructed man's mind beyond the wildest dreams of its founders. Always a vehicle of progress, the Classic had become an instrument of outright revolution. The Rome which inspired Christopher Wren was not the Rome which fired David; the Greece of Angelica Kauffmann was far different from that of Horatio Greenough; and interstellar space divided the Classic world of Jefferson from that of the dictator Napoleon.

Up to 1776, the Americans had found themselves one step removed from this constantly shifting image, too busy fighting the Indians to spend much time on all its nuances and subtle distinctions of emphasis. Until then architectural styles, like fashions in periwigs, were a luxury over which precious few colonists could dawdle. Like other merchandise, they were ordered from the mother country by the rich. From here they seeped slowly down into the living fabric of the nation's popular building. If they were workable they were adopted, modified, integrated into the prevailing idiom. But the Classic yeast was at work on this side of the Atlantic and the Declaration of Independence set it to work. Esthetic standards received a new, a political significance. As Jefferson was soon to demonstrate, it was time for Americans to create a concept of beauty peculiarly their own.

Fig. 37. Drawing room, Grosvenor Square, London, 1773. Robert Adam, architect. The impact of the study of Pompeii and Herculaneum is clearly evident in the decorative devices used here.

2. 1776–1820

THE NEW REPUBLIC RISES

Any estimate of American building since the Revolutionary War must always be modified by the special conditions of American society since that time. These conditions amount to a paradox which dates roughly from the day our source of supplies in England was cut off: two separate and distinct economies — one might almost say cultures — have been simultaneously employed in creating the nation. These two moved like two successive waves across the nation from east to west and from north to south. The first was that pre-industrial economy of the seventeenth-century settlers. Gaining a beachhead on the eastern seaboard, it moved slowly but steadily westward — clearing the forests, trapping the furs, establishing a small-scale, self-sufficient, agrarian economy. This culture employed its inherited building technology, with only such modifications as I have already described.

The second wave, that of nineteenth-century industrialism, began in New England with the War of Independence. It moved slowly at first. Then, with gathering momentum, it spread out from the metropolitan centers of the Northeast like a prairie fire. In the beginning, it merely followed the first wave along the rivers, trails and roads. Then, with the canals and the completion of the railroads in the post–Civil War period, it overtook the first wave; and finally, with the present century, passed and completely submerged it. With the opening of the Oklahoma Territory, a geographic base for a pre-industrial economy had dwindled to the backwoods, the mountains, the submarginal lands. This new industrial economy brought, as physical evidence of its conquest, a radically altered building technology — without precedent not only in America but in the world at large.

The rapidity with which this new building was invented and developed is one of the great dramas of architectural history, and necessarily dominates the panorama of the nineteenth century. For, as the century matured, the qualitative difference between the two technologies rapidly increased, so that they soon ceased to bear more than a superficial resemblance to one another. Today, like the vermiform appendix in carnivorous animals, only the nostalgic remnants of this earlier technology remain in the form of an occasional newly built "Colonial" city hall or "English" suburban home.

This is the paradox which makes any cross section of American building appear superficially as a confusing jumble of anachronism and anomaly. Because of this paradox we have antique slum sitting cheek-by-jowl with shining streamlined factory, polished city side by side with backwoods farm. The physical contrast is appalling, but intellectually the spread is even worse. Buildings, like facts, are stubborn things; many a century-old house leads a useful life today. But many a century-old building concept also lingers on, reproducing itself in new buildings which are obsolete at birth. This contradiction has been a characteristic feature of American architecture since the Revolution; and it must be constantly borne in mind if one is to understand fully the forces at play in the period which opened with Washington's inauguration.

With the close of the war, the new republic faced, not only the task of reconstructing an exhausted and practically bankrupt nation, but of simultaneously creating an entirely new political and economic apparatus commensurate with its new status. Though Europe might not yet recognize it, the country had leaped from second-rate colony to first-rate power. For the building field, as for all other branches of the economy, this posed a new set of problems. Construction had been at a virtual standstill for the duration of the war. It was, if anything, less prepared now than before for the grandiose programs which the national, state, and local governments began to set in motion.

Fig. 38. The State House, Boston, Mass., 1795–98. Charles Bulfinch, architect.

Fig. 39. The City Hall, New York, N. Y., 1803–11. Joseph Mangin and John McComb, Jr., architects.

It was entirely logical that the first big building projects after the war should be largely governmental. The sheer pressure of new governmental affairs placed intolerable strains upon available facilities. In addition, there was the natural desire to give suitable architectural expression to the intense nationalism of the period. Thus, the state of Massachusetts had the problem of housing an entire administrative apparatus, including a large new legislature. It had the equally important ideological task of housing them *appropriately,* so that the people of Boston, as of the world at large, could see the concrete affirmation of their freedom. Massachusetts moved swiftly and impressively. Already in 1795 the architect Bulfinch had laid the cornerstone of a handsome new capitol; in 1802 the state solons were settling themselves into its new-found marble majesty (Fig. 38). Nor were the other states very far behind, either in time or in number of columns and size of dome. The larger scale, shrewd foresight and ardent optimism of these new state capitols and city halls (which were everywhere abuilding as the nineteenth century opened) were but the local reflection of an even bigger project — the construction of a new capital city at Washington.

The national capital, as it stands today, is a far cry from being the sort of city envisioned by the founding fathers. It is an anomaly: a cross between pilgrimage center, company town, and hostelry for itinerants, it is one of the world's few capital cities which is not simultaneously the fulcrum of forces of the cultural, economic, and industrial life of its country. Had the intentions of Washington, Adams, and Jefferson been followed, things would have been quite otherwise, for they visualized the new city as being an authentic metropolis, like the London or Paris they so admired. Its location, as well as its physical form, was to be an expression of this fact. For Washington was placed at the intersection of two axes, both of which were crucially important features of the new republic.

The first ran North and South; and Washington's understanding of its geo-political significance is impressive. The Allegheny-Appalachian ranges formed a dangerous line of cleavage between the seaboard states and the rest of the continent. The natural slope was toward the Mississippi Valley, where the Spanish were ensconced in Louisiana and the British around the Great Lakes in Canada. Wash-

Fig. 40. View of Washington, D. C., from across the Potomac.

Fig. 41. Plan of Washington, D. C. Charles Pierre L'Enfant, architect. This plan was issued in 1800 in an effort to speed up the sale of lots to raise money for improvement of the capital city.

ington hoped to drive a canal across the mountains, connecting the Potomac and the Ohio Rivers, thereby siphoning trade and commerce into the Atlantic and binding the transmontane West to the new nation. Even before the Revolution he had written that "there is the strongest speculative proof in the world to me of the immense advantages which Virginia and Maryland might derive (and at a very small comparative expense) by making the Potomac the channel of commerce between Great Britain and the immense Territory" to the west.[1] After the Revolution, of course, the issue was much more urgent. The new city was to be the gate between settled East and still-to-be-settled West.

But another, and equally serious line of cleavage passed through the site of the proposed city: that between free-labor North and slaveowning South. This, too, Washington fully understood. He hoped to convert Virginia into an industrial state by developing her coal and iron deposits in the mountains. Canals and highways were necessary to link the mines to the smelters and forges of the seaport. Only by such industrial development could Virginia be saved from slavery since, as he put it, "the motives which predominate most in human affairs are self love and self interest." The new capital city, gateway to this proposed industrialization, would thus act as hinge between North and South. Had his advice been followed, Virginia would have become a free state; and, as Brooks Adams put it later, "with Virginia free, there could have been no Civil War."[2]

But Washington died in 1799 and his perspectives of economic development were forgotten. Charles Pierre L'Enfant, the French military engineer, had been at work on plans for the physical development of the city since 1791. The idiom he employed was the same Baroque formalism which had served absolute monarchy so well at Versailles, Potsdam, and St. Petersburg. But whereas those royal complexes revolved around a single point (the bed, body, and sexual organs of the monarch), L'Enfant's plan was an expression of the lines of force relating the three deposits of republican power — White House, Capitol, and Supreme Court. Significantly, where these axes crossed (the present site of the Washington Monument), the founders proposed a National University. From this position at the very heart of the nation it would "spread systematic ideas through all parts of the rising empire . . . [it would] fix a standard of collective thought." Thus for all its formality, the Classic idiom demonstrated once again the versatility which had made it the absolutely universal language of the pre-industrial Western world.

America moved in many other ways to concretize the program of the new republic: and each of her moves, directly or otherwise, had its impact upon the building field. Characteristic of these was the popular demand for education. As early as 1779, even before independence was assured, Jefferson had pointed out how necessary to freedom's survival was an educated public. In his magnificent proposal for Virginia, "A Bill for the More General Diffusion of Knowledge," he wrote:

> Whereas it appeareth that however certain forms of government are better calculated than others to protect individuals in the free exercise of their natural rights, and are at the same time better guarded against degeneracy, yet experience hath shown that even under the best forms those entrusted with power have, in time and by slow operations, perverted it into tyranny; and it is believed that the most effectual means of preventing this would be to illuminate . . . the minds of the people at large . . . to give them knowledge . . . of the experience of other ages and countries, [that] they may be enabled to know ambition under all its shapes and prompt to exert their natural powers to defeat it.

Here Jefferson was outlining in prescient detail a system of universal, free, and compulsory education; and though his proposals were only partially adopted in Virginia, they served as prototypes to the institutions which were rising all across the land. New colleges, academies, and schools reflected the hunger for education at every level of society. Though many of them may today appear both pedagogically and architecturally unimportant, they nevertheless were the indispensable basis for modern standards in both education and school-building design.

The closing years of the century saw many other projects of an institutional nature. One of the more active building types was — ironically enough — the prison. In 1800 New York opened its new state prison, designed by the *émigré* architect Joseph Magnin. In plan this building provided for segregation of sexes and for grouping of prisoners according to severity of crime. This represented a big advance over contemporary European practice, which threw murderer and debtor, sick and insane, all into one foul cell together. Virginia's new penitentiary went one step further (at Jefferson's suggestion) and provided for solitary confinement. And these and similar penological reforms were soon to be reflected in the design of the Eastern Penitentiary which Pennsylvania completed in 1835. This institution was so advanced for its day as to become famous — an object of careful study by the European commissions sent over to see it and a source of much satisfaction to American reformers.

As economic equilibrium returned to the rich seaport cities and the fertile river valleys, private construction revived. Amateur designer and anonymous builder began to yield to professional architect. Among the Portsmouth shipping gentry, Mr. Samuel McIntire was much the vogue; in Boston Charles Bulfinch, overestimating perhaps the strength of his popularity, went bankrupt on an elegant group of houses modeled after those at Bath in England; in New York and Philadelphia and Baltimore the names of Robert Mills and Benjamin Latrobe were becoming famous.

Yet despite all their progressive features, these buildings — from largest capitol to smallest jail — belonged more to the period which was closing than to the century which lay ahead. They were grounded upon the building technology of preindustrial society. They made no insuperable demands upon the resources of conventional structural theory or traditional technique. They were largely content with pre-Revolutionary standards in craftsmen and building material: nor were they to yield any important solutions to the problems of heating, ventilating, lighting, or acoustics. Their chief significance lay in their planning — their ability to provide for the increasing scale, complexity, and specialization taking place within the organs of society.

These buildings were not historically retrogressive — yet. For their time they were as well built, well lighted, and well heated as similar buildings elsewhere. They met the special needs of their society adequately. They were as good as they had to be. But they were not in themselves to be the instrument of future technological advance. For this purpose, nineteenth-century industrialism was to employ an entirely new building type: the factory.

Fig. 42. Christ Church, Lancaster, Mass., 1816. Charles Bulfinch, architect.

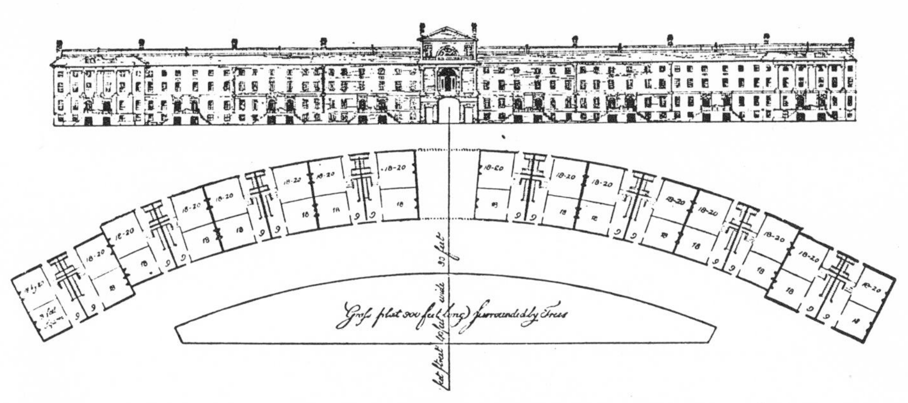

Fig. 43. Tontine Crescent, Boston, Mass., 1793–95. Charles Bulfinch, architect.

Fig. 44. Jacob Ford House, Morristown, N. J., 1774. Washington's headquarters in winter of 1779–80.

Fig. 45. View of Savannah, Ga. This print of 1855 gives a clear idea of Oglethorpe's plan of 1733.

Fig. 46. Louisiana State Bank, New Orleans, La., 1818. Benjamin H. Latrobe, architect.

Fig. 47. Plantation house, Chalmette, La., c. 1820.

Fig. 48. Jackson Square, New Orleans, La. Cabildo (left) 1795; Cathedral of St. Louis, 1794; and Presbytere (far right), 1795–1813, seen from the balcony of one of the Pontalba buildings.

PARTHENOGENESIS IN THE BUILDING FIELD

From the moment it appeared as a distinct type, with its own special characteristics, the factory was to have a twofold significance for American building of every category. Ultimately nothing would escape its impact. On the one hand, it was to supply those improved materials, equipments, and concepts without which advance in the field would have been inconceivable. (Imagine the modern skyscraper, its steel, aluminum, plywood, conditioned air, and fluorescent lighting, without the factories to turn out such products.) On the other hand, as a special type of building, the factory itself soon became the pacemaker relative to which all other types were laggard. The manufacturer demanded infinitely better performance of his building than the homeowner did of his. (It was entirely feasible to raise an entire family in a wigwam; but try to turn out a single yard of cheap printed calico under the same conditions.) To make industrial processes at all possible, the factory had to perform better than building had ever done before — that is, offer a greatly expanded and far more precise control of all environmental factors. The factory was thus at once the parent and the firstborn of modern building technology.

The beginnings were modest. Gristmills, smithies, and cobbler shops had been adequate for pre-Revolutionary handicraft production. As long as he stuck to silver (with an occasional dental plate now and then), Paul Revere could carry on in his own house, adding a room here or a shed there. It was only when he was commissioned to sheathe the new frigate *Constitution* in copper that he was forced to seek larger quarters. The level of his production had risen so sharply that the whole character of the process had been changed. As early as 1793, Samuel Slater had built a new spinning mill in Pawtucket, Rhode Island. Here water-powered machines are said to have initiated American mass production. Mr. Slater's mill was indubitably a factory: regular fenestration, angular roof, utter absence of embellishment or whimsy prevented it from being mistaken for anything else. At the same time, its clapboard walls, inadequate many-paned windows, and shingle roofs came from a long line of village gristmills. This was but one of the many structures

Fig. 49. Samuel Slater's Spinning Mill, Pawtucket, R. I., begun in 1793.

Fig. 50. Public utilities at Philadelphia, Pa.; Bridge, 1812; Waterworks, 1819–22; Canal, 1825.

Fig. 51. Woolen mill, Allendale, R. I., c. 1822. Zachariah Allen, designer.

with which New England was bidding for the industrial leadership of the nation. They spread with amazing rapidity. By 1820 they were to be found wherever waterpower could be exploited.

The rapidly increasing demands of industry accelerated the development of new structural concepts and systems. Disastrous fires and structural collapses because of overloading egged on the factory designers. An entirely new scale was given to industrial operations by Francis Cabot Lowell when, in 1814, "he revolutionized the textile industry by introducing his power loom and by organizing, for the first time, all cloth-making processes within one mill."[3] Greater spans, greater floor strength for constantly heavier machinery demanded better building materials; increased fire resistance, better heating, artificial lighting were obviously imperative (Figs. 102, 136).

The first steam-powered sawmill in the country began operations in 1803 at New Orleans, soon to become the collection point for the fabulous stands of southern pine, cypress, and yellow poplar of the Mississippi Basin.[4] By the 1820's, steam power and circular saw had put the production of lumber on a modern production basis. Wood had always been the country's favorite building material and the power-driven swamill clinched its ascendancy (in 1960, 85 per cent of the nation's houses were still of wood). Though lumber in unprecedented sizes and lengths was now plentiful and cheap, the appearance of the fully rationalized wooden-framed structure awaited another industrial component — the machine-made nail. But long before the factory system was able to supply machine-made components, the wooden frame was being rationalized. There are records of prefabricated house frames being made as early as 1578. The Louisiana French were shipping them to the West Indies as early as 1727. A prefabricated house was stolen from a Natchez wharf in 1791; and the Utopian community of the Rappists at Economy, Pennsylvania, would erect two hundred identical dwelling houses in the years 1824–1830, prefabricating them from a single set of shop drawings.

The manufacture of all the elements required for a metal skeleton was to take much longer. Although there had been small iron foundries before the Revolution, especially in New Jersey and Pennsylvania, the metal was far too costly for building. A modern metallurgical industry awaited the simultaneous exploitation of coal; and this began in earnest only in the mid-nineteenth century.

Needless to say, mass production was no guarantee of good or beautiful buildings. History is never so simple — the factory system was to produce unparalleled ugliness and squalor. Nevertheless, all major advances in building technology were to be based upon the factory; and, as a new building type, the factory was often to establish new standards for the building field.

On Washington's birthday in 1817, Philadelphia's elite are reported to have danced to the light of two thousand candles — which reminds us that neither oil lamps nor glass chimneys were yet in common use. Mr. Franklin's theory on the beneficial effects of the tub bath — as well as his shoe-shaped bathtub — was still a controversial issue; and in the Capitol in Washington, the architect Latrobe was installing the first stove. The original fireplaces had proved inadequate and he was busy in 1804 with an ingenious contraption of his own design — start of a century-and-a-half effort to keep the legislators comfortable.

JEFFERSON, GOOD GENIE OF AMERICAN BUILDING

To a far greater extent than is visible on the surface of its history, Thomas Jefferson was the guiding spirit of American building during these formative years of the new Republic. He exercised his great influence through all sorts of channels, direct and indirect. By example, as in the buildings he himself designed — his home at Monticello, the University of Virginia. By virtue of official position, as in the competition for the Capitol while he was Secretary of State and in the L'Enfant plan for Washington while he was President. By persuasive argument, as when he sold the state of Virginia on the idea of duplicating a Roman temple in Nîmes for the new State House. Finally, by proxy, as in his long sponsorship of the architect Benjamin Latrobe. Through these channels he was largely responsible for fixing the attention of the American people directly upon the cultures and buildings of Rome and Greece.

With his enormous and well-articulated comprehension, Jefferson devoted neither more nor less of his time to building design than to any of a dozen other pursuits. But when he turned to it, he saw it whole, with passion, clarity, and optimism. He saw architecture in both its theoretical and practical aspects and he understood — as few of his contemporaries did — the interpenetration of the two. Equally prepared to teach a slave to make brick, to lay out a college campus, or to design an entire architectural curriculum, he moved with ease and sureness at each level. He was as much interested in penological reform as in the plans for a new prison which expressed it; and he was more interested in an independent national esthetic standard than in any isolated aspect of our building. As much a connoisseur in the fine arts as any London or Paris dandy, he never for a moment let this polish blind him to buildings of the poor, whether in Europe or at home. And history was to prove him, on all these counts, far more perspicacious than his contemporaries.

In the design of actual buildings, Jefferson's intense practicality matched that of Franklin. Like the Pennsylvania sage, his diaries abound in keen observations on how to build better: an ingenious design for a folding table, a lamp in a fan-shaped transom which lights both sides of the door, horizontally pivoted windows which when opened would "admit air and not rain." But Jefferson's range of interests were not, like Franklin's, limited by an almost smug immediacy. In Nîmes he could sit "gazing whole hours at the Maison Carrée, like a lover at his mistress." In Holland he was entertained at a great house whose Classic façade he found "capricious yet pleasing," while in Germany "at Williamsbath there was a ruin which was clever." He was more familiar with the ruinous slums of Europe than most gentlemen, yet he could still view as "most noble" the old château above Heidelberg.

The design for the University at Charlottesville is an example of Jefferson's intelligent subordination of technique to the larger demands of theory. Here he might teach Negro slaves to carve a Classic column — changing the order to Ionic "on account of the difficulty of the Corinthian capitals." Or, in a serpentine wall he might, by canny use of curves, make one brick do the work of two. But long before he had outlined the functions of a state-wide system

Fig. 52. University of Virginia, Charlottesville, Va., begun 1817. Thomas Jefferson, architect. 1856 print shows dormitory added to rear of Library after his death and destroyed by fire in 1895.

Fig. 53. Plan of the University shows oval rooms at ground level of Library (1). Ten pavilions face lawn (2), with classrooms below and professors' quarters above, are flanked by students' rooms (3). Gardens (4) also serve six dining halls (5) and additional students' quarters in outer range.

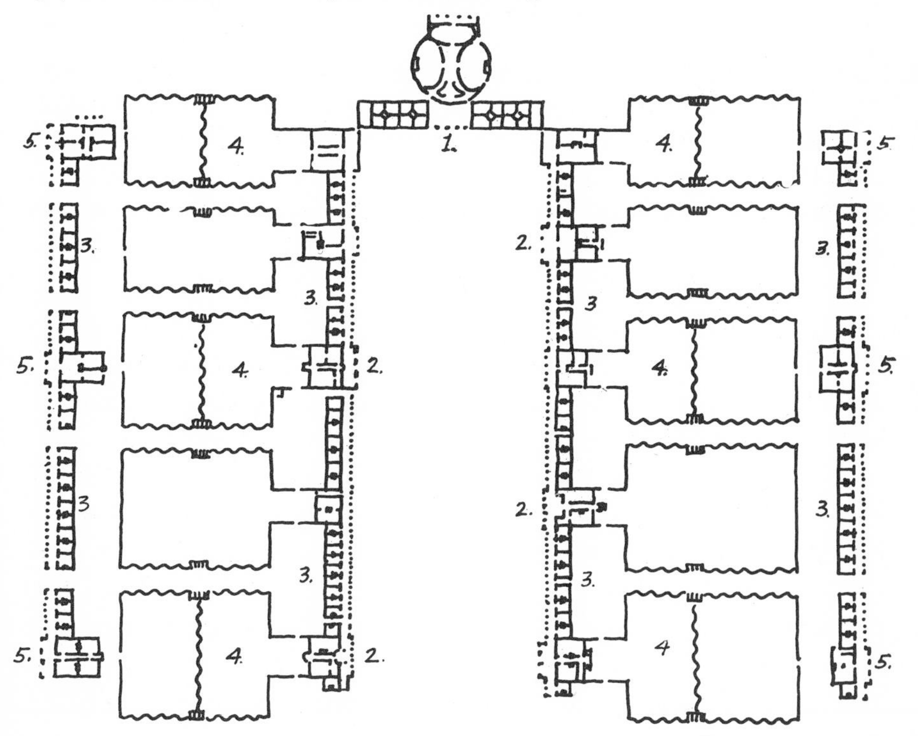

of education; and while he watched the masons at work, he had a very precise picture of the capstone of such a system.

> I consider the common plan followed in this country — but not in others — of making one large and expensive building as unfortunately erroneous. It is infinitely better to erect a small and separate lodge for each separate professorship, with only a hall below for his class and two chambers above for himself; joining these lodges to barracks for a certain portion of the students by a covered [passage] way to give a dry communication between all schools. The whole of these, arranged around an open square of trees and grass, would make it what it should be in fact — an academical village. . . . [Such a plan] would afford that quiet retirement so friendly to study and lessen the danger of fire, infection, and tumult.[5]

His approach here is rational, his concept functional, his program ambitious. He is not laying the cornerstone of some jerkwater college or rustic academy. He envisions a great university made up of many schools, designed to meet the expanding needs of a prosperous society. No abstractions, either pedagogical or architectural, are here to haunt him: he views the design problem from the standpoint of the ultimate consumers — teaching staff and student body. Each school must be separate, the teacher must have privacy, the student quiet, circulation must be convenient. His argument for the plan is not its symmetry or its style but its efficiency. It will reduce the danger of fire and disease; it will make study easier for the students and discipline easier for the teachers (Fig. 53).

Like all educated men of the period, Jefferson moved in the lustrous dawn of scientific discovery, before the disciplines had congealed into specialized compartments. A smattering of natural philosophy and an inquisitive turn of mind made any man a scientist, for the line between amateur and professional was indistinct. There was nothing of the dilettante in Jefferson's excursions into science, however. His interest in the weather, for instance, did not stop with the amusing windvane and register which he built into Monticello, nor with the records of temperature and rainfall which he kept for many years. The relation between climate and comfort, and the role of building in modifying this relationship, intrigued him. He felt that the standards of construction were far short of what they should have been.

> Private buildings are very rarely constructed of stone or brick, much the greater portion being of scantling and boards, plastered with lime [that is, conventional clapboard houses plastered inside]. It is impossible to devise things more ugly, uncomfortable, and — happily — more perishable. There are two or three plans on which, according to its size, most of the houses in the state are built. The poorest build huts of logs, laid horizontally in pens, stopping the interstices with mud. These are warmer in winter and cooler in summer than the more expensive construction of scantling and plank.[6]

From a structural point of view, he was wrong about the frame house: as a piece of engineering it was remarkably efficient. Nor was it as short-lived as he assumed, thanks to the rapidly growing use of oil paints. But he was entirely correct in observing that frame construction offered a much less efficient barrier to the transmission of heat than either log or masonry. The colonists' prejudice against masonry houses (aside from their greater cost) was largely based upon the condensation which occurs on masonry walls under certain atmospheric conditions. This "weeping" had given birth to wild tales of disease and death caused by malignant vapors. Calmly (in what was perhaps the first explanation of its kind in history) Jefferson undertook to show that nothing more dangerous than water vapor was involved, to explain the phenomenon clearly and precisely, and to outline measures which would prevent or minimize its occurrence.

Chronologically, Jefferson's understanding of the importance of a national architecture steadily deepened and matured. He was a critic of taste, perceptive but subjective, when in 1782 he penned his scathing indictment of Williamsburg architecture. He was, of course, born a Classicist. But there was room for sharp disagreement within the movement; and when he attacks "the barbarous burthen" of Georgian ornament, he is already identifying himself with the radicals of the nascent Classic Revival. Federalist decor was a pallid and upper-class nationalism; its mere substitution of the American eagle for the profile of the British king was not sufficient. In all

Virginia, he found only four public buildings worthy of note — the capitol, the palace, the college, and the insane asylum, all of them in Williamsburg. The capitol was, he felt, "a light and airy structure . . . on the whole the most pleasing piece of architecture we have." The palace was "not handsome without, but spacious and commodious within." As for the college and asylum — they were "rude mis-shapen piles which, but for the fact that they have roofs, would be mistaken for brick-kilns" (Fig. 23).

All in all, it seemed to him that "the genius of architecture had shed its maledictions over this land."

> To give these buildings symmetry and taste would not increase their cost. It would only change the form and combination of the members. This would often cost less than the burthen of barbarous ornament with which these buildings are sometimes charged. But the first principles of the art are unknown, and there exists scarcely a model among us sufficiently chaste to give an idea of them . . . perhaps a spark may fall on some young subjects of natural taste, kindle up their genius, and produce a reformation in this elegant and useful art.[7]

Here his chaste model is obviously that of ancient Rome as against that of Georgian London. There is more than a little political animus in his position, which was natural enough, with the Revolution only now being brought to a successful conclusion. His preferences were all with "columnar" architecture (as the Revival styles were then called), but these preferences were not yet dominantly ideological.

He proposed columns because he liked them, and brick because "when buildings are of durable materials, every new edifice is an actual and permanent acquisition to the state, adding to its value as well as to its ornament." Here he sounds a new and significant note, the patriot's desire to see his country beautiful, wealthy, strong, and respected. As a matter of fact, this sturdy concern for the social wealth of the nation was to be a substructure for his lifelong attitude toward architecture. He wanted national greatness at every level of national life — political, social, cultural, artistic. The building field, like any other, must contribute to this greatness; and the individual building itself must satisfy, not only the needs of its owner, but also of the community as a whole.

Fig. 54. Virginia State Capitol, Richmond. In 1786 Jefferson sent plans and plaster model from Paris. Using the Maison Carrée as a model, he changed columns from Corinthian to Ionic to simplify construction.

Fig. 55. Maison Carrée, Nîmes, France, 16 B.C. Thomas Jefferson sat before this Augustan temple "like a lover before his mistress," mistaking its imperial splendor for that of republican Rome.

But above and beyond its obvious utility, Jefferson saw great architecture as itself a civilizing force. That is why, at the University in Charlottesville, he designed each of the pavilions in a different variant of the Classic orders. They were to act as "models of taste and good architecture and [to be] of a variety of appearance, no two alike, so as to serve as specimens for architectural lectures."[8] Thus he proposed to indoctrinate *all* the sons of the rough Virginia gentry (and not merely the aspiring architects) with at least the minimum of artistic literacy. The very buildings, like the classes held inside them, would help to prepare the students for their role of leadership.

Jefferson's advice to American tourists in Europe is a classic of shrewd selection, written by a very Baedeker of social acumen. It might pay them to have a look at the courts of European royalty, much as one might go to the circus, remembering always that "under the most imposing exterior, [courts] are the weakest and worst parts of mankind." But the items of real importance to American visitors were six:

> 1. Agriculture. Everything belonging to this art . . .
> 2. Mechanical arts, so far as they respect things necessary in America, and inconvenient to be transported there readymade, such as forges, stone quarries, boats, bridges (very especially) . . .
> 3. Lighter mechanical arts and manufactures. Some of these will be worth a superficial view; but . . . it would be a waste of attention to examine these minutely.
> 4. Gardens. Particularly worth the attention of an American.
> 5. Architecture worth great attention. As we double our numbers every twenty years, we must double our houses. . . . Architecture is among the most important arts; and it is desirable to introduce taste into an art which shows so much. . . .
> 6. Painting. Statuary. Too expensive for the state of wealth among us . . . worth seeing but not studying.[9]

This is the advice of a realist, neither snob nor hayseed. Europe had much to teach us. There were many interesting areas of activity over there and ultimately Americans must master them all. But time was short. There was a nation to build and our scale of values must be realistic. Painting and statuary might be too expensive — at least for the present. Architecture was a different case. Jefferson's dis-

Fig. 56. Monticello, near Charlottesville, Va., 1769–1809. Jefferson worked on his beloved home for forty years, extending and remodelling it but always holding its essentially Palladian plan.

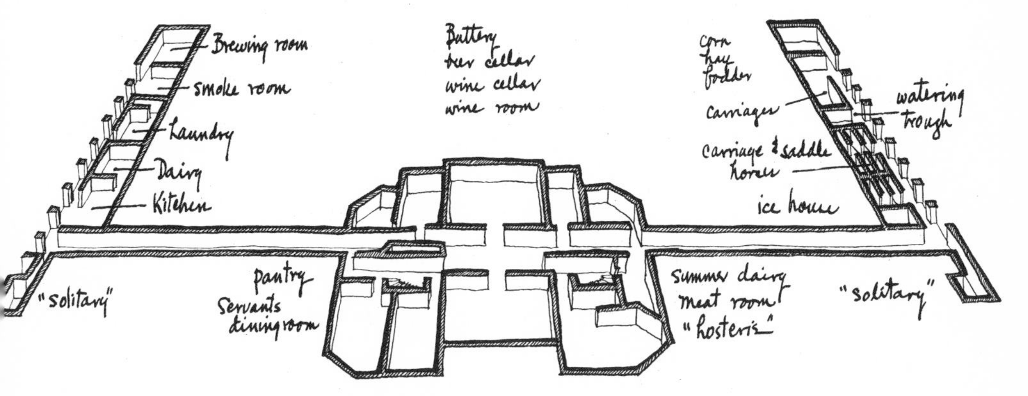

Fig. 57. Ground floor plan Monticello. By placing all service elements one level below the *piano nobile,* Jefferson gave house servants all-weather circulation. Two enclosed stairways gave separate access to all floors, thus isolating "served" from "serving" areas completely.

trust of the aristocracy extended to the palaces which housed it. His notes on his tour of the great English estates are almost contemptuous of the houses themselves, critical of their gardens. Nonetheless, Europe must be studied, the wheat separated from the chaff. Since there was undoubtedly a lot of building to be done in the new Republic, he was determined that it be sound, because that would increase the social wealth. He also wanted it to be beautiful because *it shows so much* — that is, the world would see our building and judge us by it.

THE ROMAN IDIOM: INSTRUMENT OF PROGRESS

As Jefferson's attitude toward architecture was maturing, he began to see in it an important instrument of ideological persuasion. This concept was clearly related to his appreciation of the situation in which the country found itself. Faced with enormous responsibilities — mistress of a continent, her population small, her productive capacity weak — she was strong only in potentials. No one understood this dialectic of strength and weakness better than Jefferson. America must have time to grow and room to expand in; and no stratagem which held at bay the arrogant nations of the old world was too small to be overlooked. A handsome capital city in which to receive foreign diplomats decently — this was a real necessity in the tricky, shark-filled shoals of international diplomacy. A "tasteful" dinner party in well-designed and well-furnished rooms would serve to blur, if not to conceal, the complete absence of a naval fleet. A lavish table with choice wines would contradict ugly rumors of bankruptcy; and public buildings like the Classic State House at Richmond would partially correct the rude impact of the log cabins in the pinelands.

If years in public life had given Jefferson this attitude toward architecture, then his long stay in France wedded him to a specific idiom: the Greco-Roman "columnar" style. He had long been partial to it. Now personal inclination deepened into political conviction. In the French Revolution he saw the logical extension of the American War of Independence. A close and sympathetic observer, he learned much that was to stand him in good stead when, in his battles with the Federalists a decade or so later, he was to fight for and win similar extensions of democracy for the common people of his own country. He learned, among other things, the strategic importance of the fine arts. In Tom Paine's pamphlets he had had conclusive proof of the political value of writers to the revolutionary cause. Now, in France, he saw all the arts mobilized in the democratic struggle. Painters, sculptors, writers, and architects were not only organized, their production put to explicitly ideological uses; they were themselves in the forefront of the struggle. Jacques David, official artist of the Revolution, did not content himself with heroic canvases. He participated in the drafting of the Constitution of 1791. Later, when the king tried to negate its moderate provisions, he accepted the presidency of the revolutionary Convention. The whole ponderous machinery of the Academy had begun to turn faster in the service of the revolutionary government. In return, for the first time in modern history, the government had voted the artist a living wage.

And what was the language that these radicals spoke? David, the leader, turned to the Roman Republic for his idiom. He cast his characters in the roles of Roman allegory, clothed them in Roman togas, housed them in Roman buildings, furnished them with Roman beds and chairs (Figs. 60–62). Others followed suit. Thus David shared with Angelica Kauffmann this historical necessity: neither could escape the reference frame of all European culture, both must return to the Classic fountainhead of Greco-Rome for nourishment and renewal. But between Kauffmann's oversexed and misty cherubs and the lucid, sharp-edged reality of David's perspectives lay all the vast distance of their social and political divergence. She painted for the pompous, powdered British bourgeoisie, already fat with a century of power. He spoke for forces which were only now challenging the absolutism of the French kings. And, under the impact of tumultuous events, the distance between the two was constantly increasing, though they both revolved around the same axis.

But Jacques David's Socrates is actually drinking the hemlock in a Roman, not a Greek, prison. Here, as in many of his other paintings of the period, David's architectural settings are Roman though

Fig. 58. Ceiling medallion in boudoir, 20 Portman Square, London, 1773. Angelica Kauffmann, painter.

Fig. 59. Decorations, Syon House, near London, 1765–73. Angelica Kauffmann, painter.

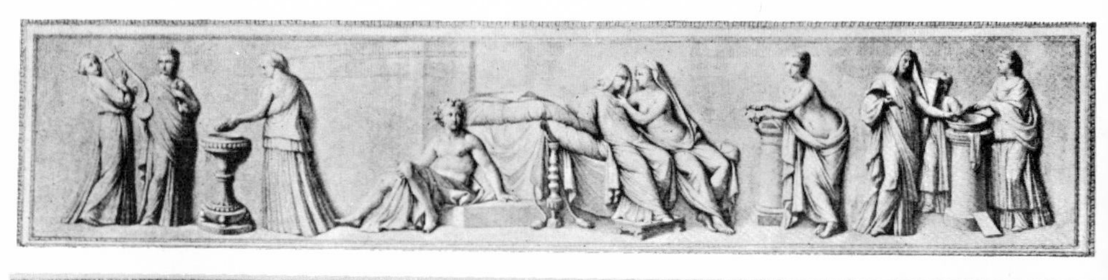

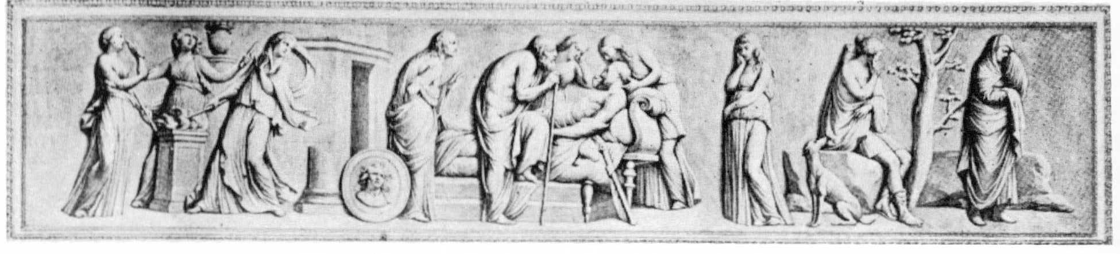

Fig. 60. "Portrait of Mme. Récamier," 1800, by Jacques Louis David.

Fig. 61. "Le Serment du Jeu de Paume," 1790, by Jacques Louis David. In best Neo-classic tradition, first studies for this painting were done in the nude (below).

his subject matter may be Greek. And this painted architecture closely resembles the work of two of his contemporaries, the architects Claude Nicholas Ledoux and Étienne Louis Boullée. Both of these men were products of the Enlightenment: they had rejected the whole body of Renaissance-Baroque tradition and returned to the fountainhead of Classic antiquity. In a series of remarkable projects (none of which was ever built) we see an architecture of true reason. Here is Roman architecture stripped of all localism and mythic anecdote, distilled into an architecture of pure geometry: cube, pyramid, sphere, and cylinder. In their noble gravity, these projects clearly outran the technological possibilities of their day but, in retrospect, they seem to have lacked only reinforced concrete to have made the Classic once again a viable idiom (Fig. 63).

There is no evidence that Jefferson ever met any of these men but he made it his business to keep abreast of all current work in architecture and art. Thus, in 1787, he wrote with admiration of current improvements then under way in the French capital, including Ledoux's designs for "the wall of circumvallation round Paris and the palaces [i.e., the combination guard and gate houses] by which we are to be let out and in."[10]

What was the attraction which pulled the French artists ever closer to the ancients? Tom Paine, who was always punctual in his appointments with history, showed split-second timing again:

> It is owing to the long interregnum of science, *and to no other cause,* that we have now to look through a vast chasm of many hundred years to the respectable characters we call the Ancients. Had the progression of knowledge gone on proportionably . . . those ancients we now so much admire would have appeared respectably in the background of the scene. But the Christian system laid all waste; and if we take our stand about the beginning of the Sixteenth Century, we look back through a long chasm to the times of the ancients, as over a vast sandy desert, in which not a shrub appears to intercept the vision of the fertile hills beyond.[11]

It was nothing less than the historical continuity of intellectual progress which was at stake in France. The beady eyes and slavering tongues of Europe's blackest reaction ringed her round like wolves in the forest. Every weapon was needed and the artists did not fail her. On the ideological level, David's message was the same as Paine's; only their media differed.

Fig. 62. "La Mort de Socrates," 1787, by Jacques Louis David.

Fig. 63. Cross-section, Opera House (project), 1789. Charles Étienne Boullée, architect. One of many Utopian structures visualized by Boullée as the architectural setting for the new regime of reason.

That David, adopting the Roman idiom, was guilty of artistic anachronism is beside the point. The entire Classic movement — from Inigo Jones to McKim, Mead, and White — was by definition guilty of archeological distortion. These discrepancies are apparent only *in retrospect:* for, as Milton Brown has pointed out, the artist was not the only one who was guilty of historical bias or architectural distortion:

> His contemporaries, through a similar association, had also acquired these same delusions, with the consequence that there was no fallacy [apparent to them]. It is evident, therefore, that the revolutionary approach to antiquity was no longer one of archeological or esthetic interest, but was one of social and political significance. Antiquity was valued because in it resided the virtues most desirable to the contemporary mind. Political republicanism logically gave rise to artistic republicanism.[12]

Thomas Jefferson was by no means an unwitting victim of this ideological mechanism. On the contrary, he understood it quite clearly. Like Paine, he saw in the theocratic Middle Ages nothing but bigotry, reaction, material retrogression. The Ancients were his friends and counselors. Had he not, when they wrote him from Virginia for an original design for the new capitol, sent them instead the measured drawings of the Maison Carrée in Nîmes? For him the request had been "a favourable opportunity of introducing into the State . . . this most perfect model" of Roman architecture (Fig. 55). He was already fully prepared emotionally to accept the new idiom: his travels along the Mediterranean littoral, his moonlight visits to the Forum, his contacts with Winckelmann's researches at the Vatican — these served to change preference into conviction.

In March, 1789, the capital of France was full of art. But Jefferson, making the rounds of the galleries, wrote Madame de Bréhan: "I do not feel an interest in any pencil [i.e., brush] but that of David." His unfailing political acumen was a sure base for his esthetic standards. The art of Europe *in general* was "too expensive" a luxury for Americans, worth "seeing but not studying." This dictum obviously did not include the productions of David and his colleagues, for the good and sufficient reason that they had a practical significance to

American problems. Just as the French intelligentsia recognized in Jefferson the most advanced exponent of American democracy, so he recognized in them the greater political maturity of the French Revolution. One of the evidences of this maturity was the use they made of art, the effectiveness with which they had mobilized the artists in the people's behalf, and the rich and productive idiom they had chosen in the Roman Revival.

Needless to say, the Classic Revival in this country was not the subversive importation of Jefferson alone — though there were those who said it was. On the contrary, as a movement it would never have got to first base had not its symbolism corresponded to the needs of the dominant classes of the period and been consistent with their general outlook on the world. For, with the final adoption of the Bill of Rights, the state power had been wrested, by those forces which had fought the War of Independence through to a successful conclusion, from the hands of those Southern slaveholders and Northern Tories who sought to hold it. Only then was bourgeois democracy, of which Jefferson was the first great spokesman, complete and secure. The aspirations of this society were rational, expansive, optimistic, breathing confidence in Man, his native goodness and nobility, his natural rights. The desire for a truly national culture based upon such concepts was a quite explicit factor of the period. In attempting to express these concepts, to concretize them into esthetic standards, it was inevitable that the nation impress them upon its building. In the context of the period, it was inevitable that the architectural language of these standards be that of the Roman Republic.

The esthetic unity of this period, so much remarked by historians and lamented by nostalgic critics, was in fact due to the very structure of Jeffersonian society. The dominant classes — the new capitalists and manufacturers, the small-propertied people, the independent farmers — were united upon a program of national reconstruction and expansion. And, so far from being a foreign "ism," Jefferson's leadership suited them to the extent that, in 1804, they returned him to the White House with 162 out of 174 electoral votes. In the building field specifically, his Classicism did not yet run counter to technological development. Moreover, ownership being very widely dispersed in such a society, a far larger proportion of the population was able to participate directly in the building process. With comparatively little centralization of landownership and none at all of construction and design, democracy in the establishment of both esthetic and technical standards was entirely logical.

For the rest of his life after his retirement from the presidency, Thomas Jefferson was to occupy a strategic position with reference to the main lines of development in American building. And his efforts in behalf of a national architecture and a native building technology were not wasted. America built hugely and built well; and the Revival — first the Roman, then the Greek — became the absolutely universal idiom of design. Under Jefferson's pervasive and kindly genius, a whole school of great American architects appeared: Latrobe, Magnin, Mills, Strickland. These men differed from their predecessors, and indeed from Jefferson himself, in several important respects. They were not amateurs, not dilettantes, not even necessarily gentlemen, but full-time technically trained professionals. They were largely the product of European universities (despite Jefferson's plans, it was not until 1866 that this country saw its first architectural school), with a sound footing in engineering and a diminished obsession with the more literary aspects of architecture. The environment in which they matured, and which Jefferson had done so much to guarantee, was especially favorable to the development of well-rounded designers. On the one hand, the emphasis of all higher education upon what would today be called the liberal arts was still large. On the other hand, the Industrial Revolution gave increasing importance to science and technology inside the classroom and out. Classicism could thus coexist with nascent science. One man could master both architecture and engineering. Both fields remained comparatively simple; the split between them was still a problem of the future.

For men like Benjamin Latrobe, it was thus possible to be at once an enthusiastic and meticulous Classicist and an imaginative and resourceful hydraulic engineer. He had landed in Norfolk in 1796, with no capital but an engineering degree from a German university and an apprenticeship under the English architect Samuel Pepys Cockerell, one of the pioneers of the Greek Revival. A versatile young man, Latrobe had spent the first few months in this country "idly engaged [in] designing a staircase for Mrs. A.'s house, a house and office for Captain P., tuning a pianoforte for Mrs. W., scribbling doggerel for Mrs. A., tragedy for her mother and Italian songs for Mr. T."[13] This lapdog existence did not content him for long. His first employment was at Richmond, doing navigation work on the James River. It must have been here that he and his work first came to Jefferson's attention. In any event, two years later he was able to carry with him to Philadelphia letters of recommendation from Jefferson. Whatever the details of their meeting, the relationship between the two men was cordial, productive, and long-lived. It is easy to understand that the young man appealed to Jefferson. He had a first-rate technical education (hard to find in a native-born American); he was a "bigoted Greek" in his esthetic standards; he had a liberal and imaginative mind.

Latrobe's career, on the other hand, owed much to Jefferson's patronage. One of his important commissions — the Virginia State Penitentiary — was based upon data which Jefferson had forwarded from Europe during his ambassadorship. The Jefferson letters helped him in the Quaker City, where he was soon designing the handsome classic structure for the Bank of Pennsylvania. It was here, too, that he received what was perhaps his most notable commission — the Philadelphia Water Works. It was certainly one of the most important building projects of the new Republic (Fig. 64). Here Latrobe displayed exactly those qualities which Jefferson so admired in building design: technical progressiveness, competent craftsmanship, esthetic sophistication. There was no prototype for a project of this complexity and scale. It was, in itself, paradoxical: it straddled the two technologies of the period. In concept and function, it belonged clearly to the new industrial society. In actual design and construction, it necessarily relied upon the techniques and materials actually available. Latrobe met and — for the moment — mastered the essential contradiction between the Classic discipline and that of the new machines. To accomplish this he had to resort to certain devices which would become all too familiar

Fig. 64. Water Works, Center Square, Philadelphia, Pa., 1799. Benjamin H. Latrobe, architect.

as the century wore on. External symmetry did not always represent internal balance. Chimneys did not always leave the roof exactly over the spot where they started below. Nevertheless, his accomplishment was great, running as it did from the installation of the pumps and design of the conduit system to a façade of impeccable taste. The steam pumps were much the largest ever built in this country and the architect risked much on them. The night of January 21, 1803, found Latrobe himself on the scene, with only three of his closest friends, nervously kindling the fires which would set the ponderous machines in motion. They did not explode, as any boiler was apt to do in those days, and his reputation was secure.

President Jefferson recalled Latrobe to Washington in 1803 to take charge of the construction of the new Navy drydocks; and the next year appointed him to the specially created post of surveyor of building for the federal government. In this capacity he designed the south wing of the Capitol, until he quarreled with Dr. Thornton, winner of the national competition for the original design. Through-

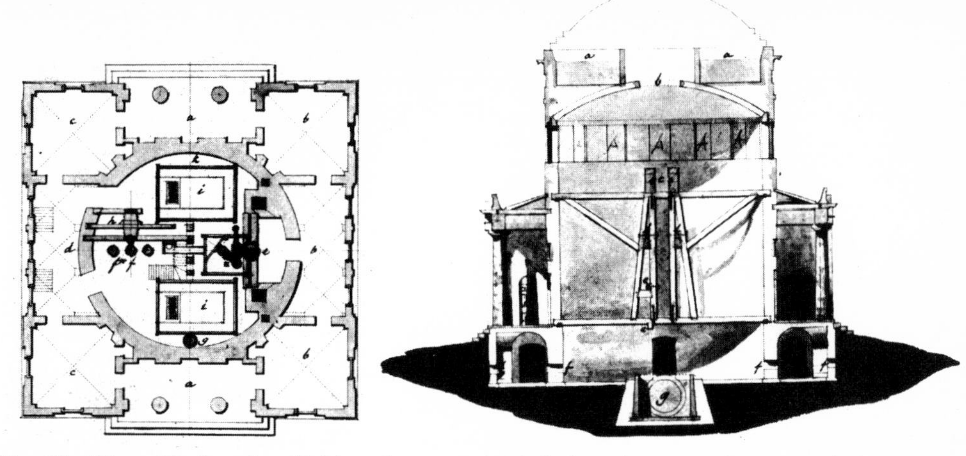

Fig. 65. Water Works, plan (left) and section (right) show how Latrobe reconciled requirements of new technology with formal criteria of Neoclassic style. Note boilers ("i") and pumps ("f").

Fig. 66. Latrobe's own drawings for another pump, dated 1808, show the refinement of line and elegant simplicity which were to make mid-century American machine design famous around the world.

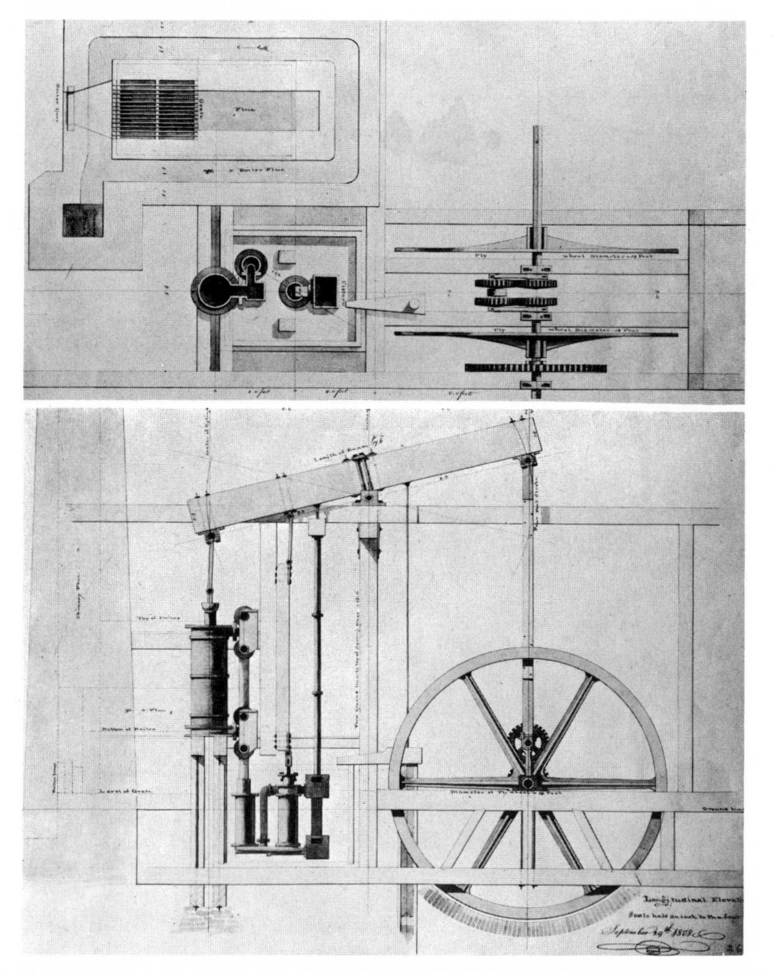

out this period Latrobe and Jefferson were closely associated. The fact that Jefferson was closely identified with the Roman, and Latrobe with the Greek, aspects of the Revival does not seem to have created any rift, despite Latrobe's somewhat didactic letter to the President declaring that his "principles of good taste were rigid in Grecian architecture." Actually, Jefferson was himself a Greek scholar — perhaps a better one than Latrobe — and the men were in agreement on their attitude toward architecture as a whole.

Besides, the aging Jefferson had lived to see the symbols of his beloved Roman appropriated by the Napoleonic dictatorship and indubitably turned from instruments of liberation into the heraldry of imperialism. The rise of the Directoire style was merely the esthetic expression of a much deeper political fact. The French Revolution had not only failed to hold many of the gains it had staked out for the common people; under Napoleon, it was being converted into a reactionary instrument which threatened all of Europe. Many of its intellectuals, including David himself, were now in the courts of the Bonapartes. Jefferson saw all this and his confidence, not in the Revolution but in the intellectuals, wavered.

He saw other things, too, which must have made him more sympathetic to younger men like Latrobe. In 1788 he had said that "circumstances render[ed] it impossible that America should become a manufacturing country during the time of any man now living." He himself had lived to see the impossible taking place. His agrarian republic was being submerged by an industrial democracy whose best exponents were men like Latrobe: Latrobe, who had designed the first drydocks in the New World, who had watched with deep interest the manufacture of his steam pumps in a Trenton factory, who was planning even then with Robert Fulton a fleet of big steamships for the upper Ohio. Jefferson was not at odds with such men; he was merely too old to keep up with them.

Practical experience in such projects as the waterworks in Philadelphia and New Orleans prevented Latrobe's getting lost in fruitless stylistic controversy. Already, before his untimely death of yellow fever in 1820, he perceived the limits of the Classic Revival — and this at the very moment it had reached its greatest peak. "Our religion requires a church wholly different from the [Greeks'] temples, our legislative assemblies and our courts of justice buildings of entirely different principles from their basilicas; and our amusements could not possibly be performed in their theaters and amphitheaters."[14] Such perspicacity must not be credited to native wit alone. More than any architect of his day, Latrobe had worked in manufacturing and engineering construction. Though its levels were not yet high enough to confront the designer with apparently insoluble contradictions, the crisis loomed on the horizon. And had he lived to face it, there is little doubt as to where his choice would have lain: Latrobe was not a man to be paralyzed by abstractions, no matter how symmetrical or chaste.

Fig. 67. Roman Catholic Cathedral, Baltimore, Md., 1806–18. Benjamin H. Latrobe, architect. Side elevation (above) and plan (below) show Latrobe's familiarity with idiom of Boullée and Ledoux.

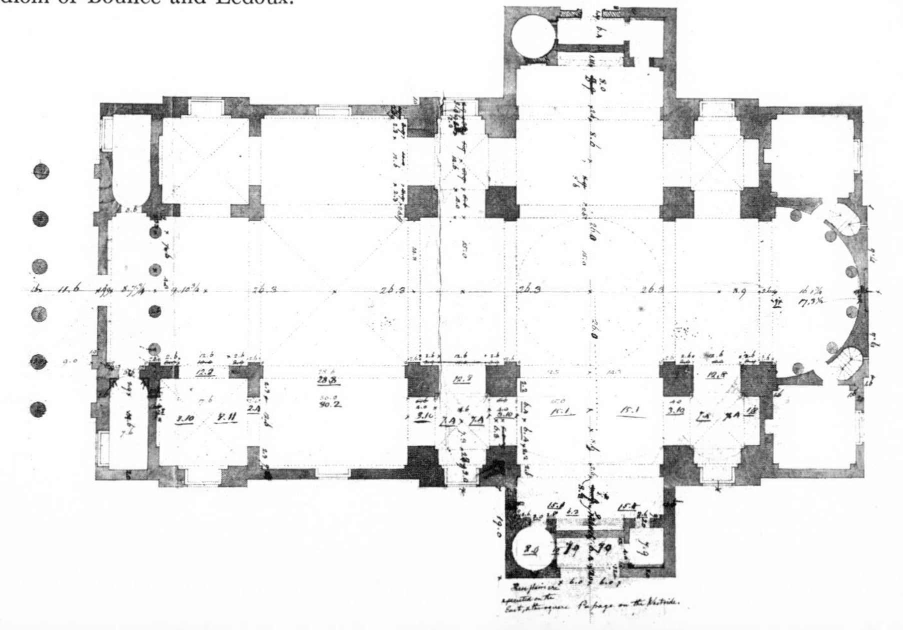

3. 1820-1840
QUIET BEFORE THE STORM

The period had opened auspiciously. The shifting relationship between agrarian and industrial economies came into a temporary national equilibrium with the election of Andrew Jackson to the presidency in 1828. For the time being, industry advanced in the North while agriculture held sway in South and West. But the nation was already half slave and half free; and ominous portents of the "irrepressible conflict" lay in the struggle around each new state admitted to the Union. Eli Whitney had unwittingly done much to hasten and deepen the split with his invention, in 1794, of the mechanical cotton gin. Instead of five pounds, a slave could now clean five hundred pounds of cotton per day. Instead of being one of the most expensive of fibers, cotton suddenly became the cheapest. The effect on American society was cataclysmic. On the one hand, Whitney's gin laid the basis for expansion to monstrous proportions of human slavery in the South. But cheap cotton in turn made possible the rapid development of the textile industry in New England, with its liberating effects, not only upon science and technology but politics and culture as well.

The development of these two conflicting cultures — slavery in the South, industrialism in the North — had a decisive effect upon American building. As a matter of fact, the Mason-Dixon Line acted like a solid Chinese Wall around the South against which all the great progressive tides of mid-century American life beat in vain. It was the expanding industrial culture of the North which gave the nation the inventors and pragmatic scientists of the period — Fitch,

Fulton, Goodyear, Otis, Morse, Bell, and a host of others. It was New England which fathered the great literary figures of the period — Emerson, Thoreau, Whitman, Whittier, Longfellow, and the pro-democratic movements for free education, woman's suffrage, trade-unionism. It is thus not accidental that the leading architects and engineers were likewise to be found in the North — Latrobe, Mills, Haviland, Isaiah Rogers, Bogardus, and Roebling.

The plantation system of production, on the other hand, had no need for advanced building types, and the slave power could not tolerate progress in building technology any more than in other areas of human affairs. It would be a mistake to think that this condition was peripheral to the slave culture. There was a certain southern journalist who in 1857 was bitterly aware of the interconnection of events. "We have got to hating everything with the prefix *free*," he said, "from free Negro to free will, free thinking, free children, and free schools — all belonging to the same brood of damnable isms."[1] A society which did not encourage a public-school system would certainly not evolve new and better school buildings. Nor would a society based on hand labor encourage the labor-saving devices of scientists, engineers, and technicians. Instead it produced the sleazy racism of Mississippi's Dr. S. A. Cartwright, with his "scientific proof" of the necessity of slavery. Dr. Cartwright claimed the discovery of a peculiar mental disease called *drapetomania* (sic!) which caused cats and slaves to run away at night. Also, he had found that the brain of the Negro froze in the cold climates north of the Ohio; common kindness dictated they be kept in the South!

A whole literature, shallow where it is not actually mendacious, has grown up in recent years around the purported "grace" and "beauty" of life on the slave-powered plantations. But one need only turn to contemporaneous accounts, by such acute observers as Benjamin Latrobe, Fredrika Bremer, or Charles Dickens, to see how disastrous was the system for black and white alike. In his famous essay on slavery,[2] Dickens quotes the advertisements for runaway slaves in southern newspapers: their brutality is stupefying. One Negro woman had on "an iron collar with one prong turned down" while another had left with "an iron bar on her right leg." Still another had escaped with her two children after the owner says he "had burnt her with a hot iron on the left side of her face. I tried to make the letter M." The men fared no better: a twelve-year old boy had escaped "wearing round his neck a chain dog-collar with 'De Lampert' engraved on it." And missing was the slave Henry with "his left eye out, some scars from a dirk . . . and much scarred with the whip." But the sheer inefficiency and squalor of slavery was also noted by outsiders, like that model of brisk New England sanity, Julia Ward Howe:

The whites stand around with their hands in their breeches pockets, and the blacks are helping them do nothing. Fences are down, doors are ajar, filth is in the streets, foul odors in the air, confusion and neglect are everywhere. Go into a house late at night, they are all lounging about, too lazy to go to bed; go in the morning, they are all yawning in bed, too lazy to get up. No one has prescribed duties — the master scolds and drives, the slave dawdles and shirks; and if anything *must* be done, it takes one white longer to hunt up two Negroes and force them to do it than it would take one uncorrupted [Yankee] to finish it alone.[3]

Twentieth century apologies for plantation life may appear to be supported by such skillful restorations as those at Williamsburg, Monticello or Mount Vernon. But as they stand today, detached from the ugly and cumbrous system which made them possible, these graceful old houses are mere relicts. A handicraft, slave powered economy was by definition backward technologically. Productivity and efficiency on the job were dependent upon literacy. But the slave who could read the carpenter's handbooks of the North could also read an Abolitionist tract. Yet there were few cities in the South large enough to support the skilled, white craftsmen necessary for the architecture and furnishings of the plantation house. Hence most artifacts — stairs, fine millwork, architectural ornament, marble mantles, furniture, and fabrics — came either from the North or from Europe. This cultural poverty was embarrassing to such Southerners as Winton Rowan Helper of North Carolina, who wrote in the year before the Civil War that

it is a fact well-known to every intelligent Southerner that we are compelled to go to the North for almost every article of utility and adornment from matches, shoe pegs and paintings up to cotton mills, steamships and statuary; that we have no foreign trade, no princely merchants nor respectable artists; that in comparison with the free states we contribute nothing to the literature, polite arts and inventions of the age.[4]

It is to the North, then, and especially to New England, that we must turn in the period which was opening.

LIKE A DREAM, THE GRECIAN VISTA . . .

The Greek Revival was the dominant architectural style of the period. Unlike the Classic of Thomas Jefferson, however, the Greek Revival was not the exclusive property of one group; nor was it always the idiom of progress. To the contrary, it was used and discarded successively by the rising mercantile-manufacturing classes of the East, by the freeholders of Trans-Appalachia, finally by the southern slaveowners. And each of these groups read into its vernacular their own particular concepts of history and used it for their own especial ends. Thus, the Greek Revival appeared first in the great humanitarian renaissance in the Boston of the 1820's, becoming for a while the very language of literature, philosophy, and art. Then it swept over the mountains into the egalitarian West of Andrew Jackson and Abraham Lincoln, where it served as the vehicle for the democratic upsurge of the rising middle class. And ultimately, turned to quite other ends than its original sponsors contemplated, it was adroitly used by Calhoun as a pedestal on which to display his monstrous "democracy of tyrants" in the South.

By 1820 the cultural center of gravity of the nation lay in Boston, as formerly it had lain in the Virginia of Washington and Jefferson. Here was intense intellectual activity. Young George Ticknor had returned to Harvard in 1819, fresh from a triumphal tour of Europe. There he had seen Greece through the eyes of the great German humanitarians. He had met Byron, studied Greek, seen the intensive archeology of the Germans. He came back to hold his classes spellbound with tales of his discovery. And he was but the first of those who made a torrent from a tendency: Edward Everett, later the nation's first orator; Jared Sparks, the historian who met Byron and studied Greek at the German universities; Samuel Gridley Howe, who returned from Greece with Byron's helmet to write his famous *Historical Sketch of the Greek Revolution.* Like a dream, the Grecian vista lay before the whole American people. The Elgin Marbles from the Parthenon were now on display in London, and pictures, casts, and lyric descriptions of them were flooding the New World.

> The children, brought up on Flaxman's outlines, knew their mythology as they knew their Bible. . . . Everyone talked mythology as everyone had begun to discuss the history of religion; and the best of the New England novels, at the turn of the Forties . . . presupposed a feeling for ancient Greece as they took for granted a circle of readers steeped in the Bible and Latin authors.[5]

The architects of the period were not less enthusiastic Greeks than the teachers, artists, and writers of New England.[6] In fact, in strict chronology, the architects had anticipated the Boston intelligentsia. At the very turn of the century, Latrobe in his Philadelphia Water Works (Figs. 64–66) and Strickland in his Bank of the United States had shown complete familiarity with the current researches into Greek architecture. There had always been a wide and eager audience for the handbooks, carpenters' guides, and architectural plates of the British publishers. As the editorial emphasis of these swung more and more toward the Greek, they carried their American audience with them. It was not long before similar works began to appear on this side of the Atlantic. Especially successful were the books of plates by the architects Asher Benjamin, *The Practice of Architecture* (1833), and Minard Lafever, *The Beauties of Modern Architecture* (1835). Both treatises extolled the Greek and contributed largely to the success of the Revival.

It must be remembered that these books were studied not by architects and designers alone; thousands of literate homeowners, intent on keeping up with the world, scanned them as anxiously as does a modern the pages of *House and Garden*. These books were especially influential because the prospective owner seldom engaged an architect if for no other reason than that there were as yet very few full-time professional designers in the country. Instead, he would show the local carpenter what plates in the book he wanted "the house to look like." This common practice led to a very wide dispersion of the Greek idiom among skilled building workers. It gave to the building of the period a remarkable homogeneity and a high degree of competence.

Fig. 68. Tremont House, Boston, Mass., 1828–29. Isaiah Rogers, architect. Plan (below) shows fully developed range of amenities from handsome public rooms to indoor baths (at rear of court).

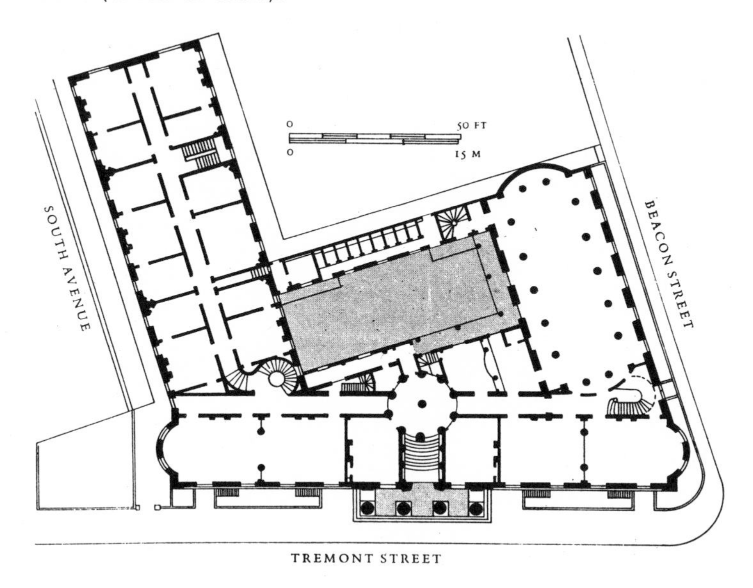

Fig. 69. Quincy Market, Boston, Mass., 1825. Alexander Parris, architect. This elaborate granite building had reception rooms and exhibition hall on second floor, double range of stalls below.

Fig. 70. Astor's Hotel (project), New York, N. Y., 1832. Town and Davis, architects. The Astor House was actually designed and built by Isaiah Rogers, the Boston architect, between 1832 and 1836.

Everywhere — in Philadelphia, in the Hudson River Valley, in Providence, Portsmouth, and Boston — Greek structures were abuilding. Thus, though Boston was the hub of the movement, its spokes radiated to every center in the Northeast and New England (Figs. 68–77). More important than its ubiquity was the fact that the Greek Revival was the idiom of the most progressive forces in American life. The chief proponents of the Greek were the very men and women who were most active in the great movements of the period. It was they who were bringing new libraries, art galleries, museums, and orchestras to America; they who were increasingly interested in the plight of the workingman, the poor, the sick, and the insane; it was they who acted as spearheads for the manhood suffrage, women's rights, and anti–child-labor movements of the period. When Mr. and Mrs. Charles Dickens came to America, it was Boston's institutions for wayward girls, orphans, and blind which most interested them. Finally — nodal point of the nineteenth century — it was the Boston intelligentsia who furnished the avant-garde in the struggle against human slavery.

Fig. 71. Design for parlors in the Greek manner, 1835. Town and Davis, architects.

Fig. 72. Girard College, Philadelphia, Pa., 1833–47. Thomas U. Walter, architect.

That the Greek Revival had the possibilities of quite other uses — as the South was shortly to demonstrate — was scarcely their fault. The fact that historically the Greek democracy had rested on the necks of foreign slaves they overlooked, much as Jefferson and the French Revolutionists had overlooked the same phenomenon with regard to the Romans. This was perhaps partly due to faulty scholarship — archeological research was only beginning to be put upon a scientific basis. It may have been partly a deliberate (though not a conscious) oversight. But it is certainly not the first example that history affords us of the use by one society of only those aspects of another which suit its particular needs.

Nor can any honest generalizations paint the town of Boston as lily-white. For she too had her Tories and appeasers, men who had little use for the democratic propaganda and even less for the social experiments which flowed from them — the great merchants who were broken to King Cotton; the narrow Royalists who had given up George III only to accept the aristocratic pretensions of Hamilton; the rising industrialists of the mill towns who opposed chattel slavery only in so far as it threatened the eighty-four-hour week of free labor. These gentry also lived in Greek houses, worshiped in Greek churches, lay buried under Greek headstones. But this was itself an evidence of the depth and vitality of the Classic movement. It was not to the Tories that the great Boston of the period belonged.

ATHENS IN TRANS-APPALACHIA

Meanwhile, west of the mountains, a new empire was abuilding. With Jackson's inauguration the middle-class frontier emerged as the driving political force of American democracy, a force which was ultimately to strike down slavery and raise New England industrialists and western farmers to power. Jackson typified this development; himself a product of the frontier, he was the first true commoner to hold the office of President of the United States. The basis had already been laid for the continental independence of the nation: the Spanish had been driven out of Florida, the French bought out of Louisiana, and Jackson himself had smashed the British at New Orleans. The way was open for the building of the West.

Fig. 73. The Hermitage, near Nashville, Tenn., 1835. Architect unknown. President Andrew Jackson built this retirement house on a fine farm. It has been completely restored to its original state.

If the West was not largely settled by New Englanders, it was at least largely to be taught by them. Throughout the territories private schools, academies, and colleges were mushrooming. The basis was being laid for that public-school system for which our Middle West is famous; and to these seats of learning flocked the Yankee schoolmarms and the Greek professors. And with them came Boston magazines, novels, and textbooks — all the cultural forms of the New England renaissance. This coincided with the political necessities of the day. As Jefferson had seen in the architecture of republican Rome the appropriate symbols for the new republic, so Jacksonian democrats saw in democratic Greece appropriate symbols for the America of the twenties. This philosophical identification was made much more real by the Greek Wars of Independence. How genuine was this sympathy, and how deep beyond the educated classes it extended, is clear both from the political rallies and money-raising campaigns to aid the living Greeks and by the adoption of the cultural forms of long-dead ones. Poems, courthouses, curricula, banks, tombstones, and dresses all bore the Grecian imprint.

Fig. 74. Map showing towns with place names of Greek and Roman origin: Aurora, Athens, Troy, Olympia, Argus, Rome, etc. Location and date of founding show how vividly Jacksonian Democracy identified itself with that of Classic antiquity.

An astonishing number of new American towns were given Greek place-names during this period; and it is significant that most of these towns lay west of the mountains — that is, precisely in that part of the country from which sprang the main sources of Jacksonian democracy. These towns had Greek architecture as well as Greek names. Ohio knew no other idiom — she leaped from log cabins to the polished severity of the Grecian idiom (Figs. 75, 77). The Ohio River was one long line of Revival towns. In St. Louis, the spreading warehouses and the new cathedral were alike in the Greek manner, while the Springfield that nourished Lincoln was a crude but confident new Athens. The Kentucky of Audubon's time and Andrew Jackson's own house (Fig. 73) were in the Greek idiom, blurred by distance and simplified by local carpenters to meet the limited materials. In the new Western style there was more than a trace of Jefferson's Rome; but nowhere at all was there anything left of the Royalist Georgian. We can only conclude then that this change in taste was in strict conformity to the qualitative changes in the structure of American society between 1790 and 1830.

But if the period closed with the Greek Revival securely entrenched as the dominant idiom of the region, it had not begun that

Fig. 75. Public Square, Cleveland, Ohio, 1839. Court House (center), First Baptist Church (right).

Fig. 76. The Indiana State Capitol, Indianapolis, 1831–35. Town and Davis, architects.

Fig. 77. The Burnet House Hotel, Cincinnati, Ohio, 1839. Isaiah Rogers, architect.

way. The settlement of the lands between the Ohio and the Mississippi was accomplished in two distinct waves, by two quite separate sorts of people. The first wave was that of the Indian fighters, the fur trappers, the "stump burners." The second was that of the farmers and town builders, the merchants and entrepreneurs who took over from the first settlers as they abandoned their clearings and moved on further into the wilderness. Each of these had its own economy, its own technology, its own building program: the log cabin was the mark of the one, the Greek Revival courthouse that of the other. As the nineteenth century approached the halfway mark, the time lag between these two waves steadily narrowed, so that in states like Indiana or Illinois they could coexist.

Life on the actual frontier, the shear line between the first American settlers and the aboriginal population, was anything but pretty. Even so enthusiastic a naturalized citizen as Hector St. John de Crèvecoeur reports that life in the great woods, in the last inhabited districts

> does not afford a very pleasing spectacle. When discord, want of unity and friendship; when either drunkenness or idleness prevail in such remote districts; contention, inactivity and wretchedness must ensue. . . . These men appear to be no better than carnivorous animals of a superior rank, living on the flesh of wild animals when they can catch them, and when they are not able, they subsist on grain.[7]

He who would wish to see America in its proper light, de Crèvecoeur warns his European visitor, must see its "feeble beginnings and barbarous rudiments" along the extended line of frontiers where the last settlers dwell:

> There, remote from the power of example and check of shame, many families exhibit the most hideous parts of our society. They are a kind of forlorn hope, preceding by ten or twelve years the most respectable army of veterans which come after them. In that space [of time], prosperity will polish some, vice and the law will drive off the rest, who uniting with others like themselves will recede still farther: making room for more industrious people, who will finish their improvements, convert the loghouse into a convenient habitation and rejoicing that the first heavy labors are finished, will change in a few years that hitherto barbarous country into a fine fertile well-regulated district.[8]

Fig. 78. Onstot Cooperage Shop, New Salem, Ill., c. 1805. This log cabin is the only original building in the restoration of Lincoln's birthplace.

Thus an urbane little city like Cincinnati could already boast (though it did not support) a cosmopolitan establishment like Mrs. Trollope's Bazaar as early as the 1820's. But scarcely a day's journey away, the Baptist missionary John Mason Peck found settlers leading a life squalid and insecure. "In and around one dirty shelter 12 feet square he found eight human beings, male and female, and the youngest full size. They were all dressed in animal skins which were covered with grease, blood and dirt."[9] They offered him pork so high he could not eat it and buttermilk from a churn so dirty he was afraid to drink it. Their house, like all the others in this landscape, was a log cabin. To build such a cabin, he tells us

> the first operation of the newly arrived emigrant is to cut about 40 logs of the proper size and length (for a single cabin) or twice that number (for a double one) . . . a large oak or other suitable timber of straight grain and free from limbs is selected for clapboards for the roof. These are four feet in length, 6 to 8 in. wide and ½ in. thick.[10]

The floors were of puncheons — wood slabs two or three inches thick, dressed on their upper surface only and laid directly upon the earth. Openings for doors, windows and chimney were cut out after the log walls were up, the severed logs being held in place by the vertical frame. There were no nails and, of course, no glass. All interstices were daubed with mud.

The furniture represented the same rough-and-ready sort of improvisation. The table was a split slab on four round legs. The bedstead was a slatted frame built into one corner. A few wooden pins set into the logs carried the women's dresses, the men's hunting shirts: a pair of buck's horns carried the rifle. A ladder led to the loft, where the children slept. Guests bedded down on the floor, feet toward the fire and fully clothed.

Cooking and eating utensils were minimal, Peck reports: "A few pewter dishes and spoons, knives and forks (for which, however, the common hunting-knife is often a substitute), tin cups for coffee or milk, a water pail and a small gourd . . . a pot and an iron Dutch oven, a tray (for wetting up corn meal for corn bread) . . . a hominy mortar and a hand mill."[11] Unbelievably crude and simple though it was, it was adequate equipment for the task at hand. And young Abe Lincoln could move directly from such a cabin to a house in town which, in terms of amenities, was literally centuries apart.

The Deep South, the South of the slaveowners, was the last area of Greek Revival penetration. This cannot be dismissed as accidental; rather, as Parrington says: "The pronounced drift of Southern thought, in the years immediately preceding the Civil War, toward the ideal of a Greek democracy . . . was no vagrant eddy but a broadening current of tendency."[12] When the nineteenth century opened, slavery was merely one of the many issues which perplexed the new Republic. Jefferson and Jackson, while disapproving of the peculiar institution, could own slaves without appearing as hypocrites. Black bodies were already a valuable commodity, but they had yet to become the indispensable base for the voracious economy of cotton. It was only as the century matured that the issues began to sharpen into first-rate political controversy until, in the fifties, it became the very fulcrum on which the fate of the nation turned. In 1830, the New England liberals could extol the Greek democracy without being troubled by its more sordid aspects. Fifteen or twenty years later, this had become completely impossible. Boston humanitarianism had moved on to Transcendentalism; Goethe and Hegel and the English Romanticists were everywhere the subject of discussion; and the Greek idiom had given way to the Gothic Revival.

In the South, on the other hand, the early enthusiasm for the Greeks had gone largely unremarked. Charleston was content with pre-Revolutionary England, New Orleans with Royalist France, as their attitudes, literature, and architecture attest. But as slavery became more important in Southern life, its defense had to be placed on a more organized basis. From apology, the South went over to attack. Searching for the best weapon, its leading ideologue, Calhoun, found the Greek Revival ready to hand where the Yankees had dropped it: "They [the Southern apologists for slavery] had been moving slowly toward the conception of a Greek democracy under the leadership of Calhoun, but now under the sharp prod of Abolitionism [publication of *Uncle Tom's Cabin* in 1851] they turned militant."[13] Calhoun spent the better part of twenty years converting the Greek Revival from an instrument of cultural liberation into a weapon of political oppression, arriving finally at his thesis of a democracy enjoyed only by tyrants. The theoreticians, professors, and architects were not far behind him. Greenough's famous article, urging Americans "to learn from the Greeks like men and not to copy them like monkeys," had appeared in the North in 1843. Not until ten years later, in August, 1853, did the *Southern Literary Messenger* get around to reprinting it.[14] Greenough himself was already dead — and the South chose blandly to ignore the very implications which he had drawn from his lifelong study of the Classic.

Charleston in the 1820's clung to her own version of galleried Georgian with its intricate detail — sometimes exquisite, often merely finicky. In the great houses which lined her streets, and the churches of St. Philip and St. Michael which towered over them, one could trace direct descent from the England of Inigo Jones, Christopher Wren, and the brothers Adam (Fig. 79). Neither Jefferson's Virginia nor the France of David had disturbed her reading habits or architectural preferences. Pope was only now yielding to Moore and the Addisonian essay. "We are decidedly more English than any other city in the United States," proudly exclaimed the Tory lawyer, Hugh Legare — the same Legare who exclaimed, "The politics of the immortal Jefferson! Pish!"; who called Jefferson "the holy father in democracy . . . the infallible, though ever-changing,

Fig. 79. St. Philip's Church, Charleston, S. C., rebuilt 1835–38 by architect Joseph Hyde. The steeple was added in 1848–50 by E. B. White. Balconied building at left is reconstructed Dock Street Theater.

Fig. 80. William Drayton house, Charleston, S. C., 1820. Like most upper-class houses, this one had large, multi-storied "galleries" oriented to take advantage of cooling breezes off the ocean.

Saint Thomas of *Canting-bury.*" No, Charleston had little use for the Greek Revival until her historic role as intellectual center of secessionism forced her to adopt Calhoun's version of it.

At the opening of the century, New Orleans — bought by Jefferson in 1803, fought over some ten years later — was still only nominally a part of the United States. Although the Yankees were beginning to come in (acting more like conquerors than newly arrived citizens, as Benjamin Latrobe tartly observed), the culture of Louisiana still derived, in about equal parts, from Spain and France. And these two powers had brought to the architecture of the region the rich experience of centuries of building in the Mediterranean, the Caribbean, and Latin America. The result was an architecture of brilliant response to environmental demands. All the characteristic features derive from the exigencies of a semi-tropic climate: the stilted pavilion form; the huge, light parasol roof and perimetric galleries; the large door and window openings, tall ceilings and through ventilation; the jalousie to solve the paradox of privacy *and* ventilation; and, in the city, the patio, loggia, and courtyard garden. All these were combined in eighteenth-century Louisiana to form the richest and most viable of all our regional architectures.

So strong and so eminently rational was this indigenous idiom that it was able to meet and absorb the ineluctable demands placed upon it by the new aristocracy of cotton. For, from 1815–1820 on, cotton replaced sugar cane and rice as the principal crop of the lower valley. The rich muck along the river was selling by the riverfront foot, like lots along Times Square. Fabulous fortunes appeared overnight in cotton. Charleston might build railroads to siphon off the cotton to the East: but New Orleans had the incomparable Mississippi. And all this swelling commerce drained into her wharves to make her the economic capital of the slave power.

New Englanders might translate Homer, Sophocles, and Aristotle, to show what heights human aspirations had reached in the past and might, with proper application, be reached again. The South would scan Plato's *Republic* for proof that slavery, and slavery alone, was the system which could make such a culture possible. Thus, if by 1850, the mid-South was studded with Greek plantation houses,

Fig. 81. Town House, Royal Street, New Orleans, La., cast iron balconies c. 1837. With high ceilings, big windows, balconies, jalousies and interior patios, these houses were ideal for the climate.

Fig. 82. Dunleith, near Natchez, Miss., 1848. A late vernacular version of the basic Louisiana prototype — parasol roof, perimetral galleries, high ceilings, floor-to-ceiling windows with jalousies.

Fig. 83. Longwood, near Natchez, Miss., begun 1860. Samuel Sloan, architect. Designed for a rich slave owner who had been to Cairo to study the growing of long-staple Egyptian cotton, this preposterous pile still retained some excellent hot-weather features, including a four-story central hall ("A" in plan) with ventilating windows in the sixteen-sided clerestory. Opposite: view of entrance front.

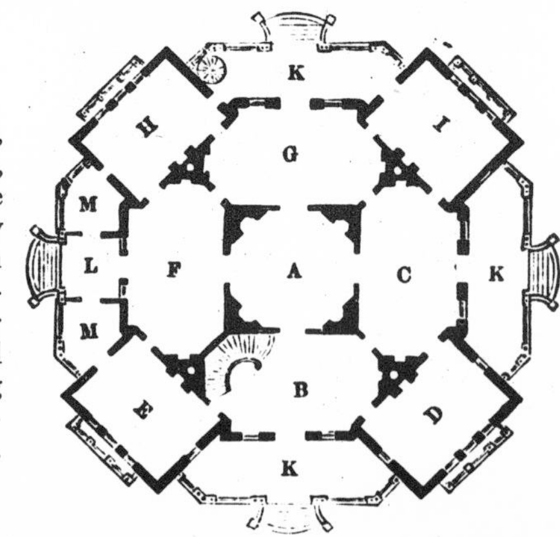

banks, and capitols, it was not the same vogue which had swept through New England thirty years before. The pedimented façade had been put to other uses. In a subtle fashion the buildings themselves reveal this. For in the South the Greek Revival was a ruling-class affair. The buildings with any pretensions to elegance, or even permanence, were those of the regnant slaveowners. Not only the slave, but the poor-white also, was pushed into the squalor of the shack; independent freeholding farmers were crushed by the pressure of slavery. In the southern countryside there was only the plantation house, the slave quarters, and the backwoods cabin. Hence there was little evidence of popular building, the anonymous structures which mark the comparatively high levels of the Ohio River valley, the Shenandoah, New York State, or the Maine seacoast.

There was, indeed, no building technology to speak of, except in the big cities. The Southern gentlemen preferred to import their skilled craftsmen, their Empire furniture and damask curtains, from New Orleans, from the North, or from Europe. Haller Nutt, of Natchez, was one such planter. From Philadelphia he had imported not only an architect but an entire crew of Irish workmen to build his house at Longwood. The Civil War broke out when the place was but half finished. As one man the crew laid down their tools, left the huge Moorish pile unfinished, and took the first boat up the river. The huge pile lies unfinished to this day. (Fig. 83b.)

A CLOUD NO BIGGER THAN YOUR HAND

What the architects of the period thought of the Greek Revival must be deduced more from their buildings than from their writings. It is true that a few of them — notably Asher Benjamin and Minard Lafever — made a comfortable living out of their vastly popular architectural books; and such books were enthusiastically Greek, as those of their predecessors had been Roman. But they involved no philosophic discussion of esthetic standards; rather, like today's women's magazines, they described how one might achieve the prevailing taste. The evidence is that the majority of architects of the period merely reflected the prevailing ideology: ready as always to trim their sails to every vagary of fashionable taste.

Like all educated men of the period, the architects were steeped in the Classic tradition, moving as freely in mythology as in the Bible. Indeed, as we have seen, the identification of past with present, of Golden Greece with Jacksonian America, was so complete, its sources so far back in our history, as seldom to call for any comment. For the building designers of that period, the Classic idiom constituted the known world. Within its limits they moved freely, with grace and knowledge. That there *were* limits, and beyond them new possibilities, seems to have occurred only to the greatest and most adventurous builders of the day.

Were these architects aware of the ideological significance of the movement of which they were a part? Nicholas Biddle, one foot in the camp of pre-Revolutionary mercantilism and one in the nascent field of modern finance, could select the Greek Revival for his own palatial residence. The financier Girard could stipulate that the Institute he was endowing be also in that style (Fig. 72). But what did Thomas U. Walter, architect for both projects, have to say on the subject? For years he had been one of the leading exponents of the Greek, evidently seeing in it the best expression of his own and his clients' aspirations. Yet by 1841 he was uncomfortably aware of the contradictions involved in the uncritical imposition of the idiom upon the reality of American life: "If architects would oftener *think* as the Greeks thought than *do* as the Greeks did, our columnar architecture would possess a higher degree of originality, and its character and expression would gradually conform to the local circumstances of the country and the republican spirit of its institutions."[15] And Robert Mills, co-designer of the Capitol and an eminently successful "Greek," warned that American architects "should never forget the original models of their country, neither its customs nor the manners of their people . . . I say to [them]: study your country's tastes and requirements and make classic ground *here* for your art!"[16] These men were mirroring the rising nationalism which was everywhere abroad in American life and letters. Be free of Europe! We are destined to lead the world, not to follow it! Walter and Mills were here not only discussing the same problem which confronted the best minds of literature, politics and philosophy; they were also discussing it at the same level and from the same standpoint — that is, as ideologues.

Among the architects of the East, then, there was by the 1840's a well-defined dissatisfaction with the Greek idiom. It was restrictive, with all the limitations of formality. Its rigid symmetry, stern affectations of simplicity, Spartan line, no longer comported well with the romantic naturalism, newfound wealth, and hearty appetites of the urban rich. As a system, it became inelastic and artificial before the complexities of industrial society. But what style would replace it? The more radical of the critics were actually attacking not merely the Greek but the entire Classic tradition.

Thomas Jefferson died on July 4, 1826 — exactly fifty years after the signing of his Declaration. But even before his death, that Classic system of which he had been the strongest exponent had come under increasing attack. Its adequacy for the new conditions of life was being challenged in England especially. There were many reasons. As the idiom of an aristocratic, pre-industrial way of life which was often rural and explicitly anti-urban, it was proving both functionally inadequate and ideologically unsuitable for the nouveau riches of industrialism. Their mills and factories, warehouses and commercial buildings were raising all sorts of new demands — greater strength, clear span and height, greater ratio of window to solid wall, etc. — which traditional structure could not meet. But in addition, with the spread of education and travel, this same new class was being exposed to a whole new range of esthetic experience. Urban wealth offered them a spectrum of possibilities which would have been inconceivable fifty years before. Against this, the cool and balanced discipline of the Classic appeared less and less adequate.

This dissatisfaction was voiced, at the critical level, by two opposing factions: the artists and intelligentsia of the radical left; and the piously conservative ideologues of the right. The point of departure in both cases was an increasing anger with the vulgarity, squalor and injustice with which rampant industrialism was clothing town and countryside alike. Radical intellectuals — men like the Englishman Robert Owen and his son Robert Dale; or Charles Fourier and his American disciple, Albert Brisbane — were confident that a bright new world could be built on the science and technology concealed in the center of this same ugly industrialism. Owen planned new cities in which "mechanism and science [would] be extensively introduced to execute all the work that is over-laborious, disagreeable or in any way injurious to human nature."[17] Brisbane's Utopian communities would be bright and airy, equipped with "convenient and labor-saving machinery . . . healthy, even elegant, workshops" which would require only "short sessions of labor."[18] None of these men seem to attach much importance to the architectural mode in which these new social values were to be celebrated. Brisbane's mentor, Fourier, proposed to house his phalansteries in Baroque palaces (Fig. 85). A drawing made for Owen shows much the same sort of palace plan but with elevations vaguely Gothic (Fig. 86). But Robert Dale Owen certainly expressed the mood of the times when he attacked the "Procrustean regularity" of the Classic and called for a functional approach. "In planning any edifice, public or private, we ought to begin *from within*," he said, before anything else considering "the specific wants and conveniences."[19]

One of the earliest men to sense the purely esthetic potentials of the industrial process had been the French architect Ledoux who, in one of his imaginary projects for an ideal town for the metallurgical industry, had proposed to transmute the very smoke, steam, and flame into ornamental features of the architecture (Fig. 84). Conservative moralists like A. W. N. Pugin and John Ruskin, on the other hand, could see no prospect of anything but disaster in the

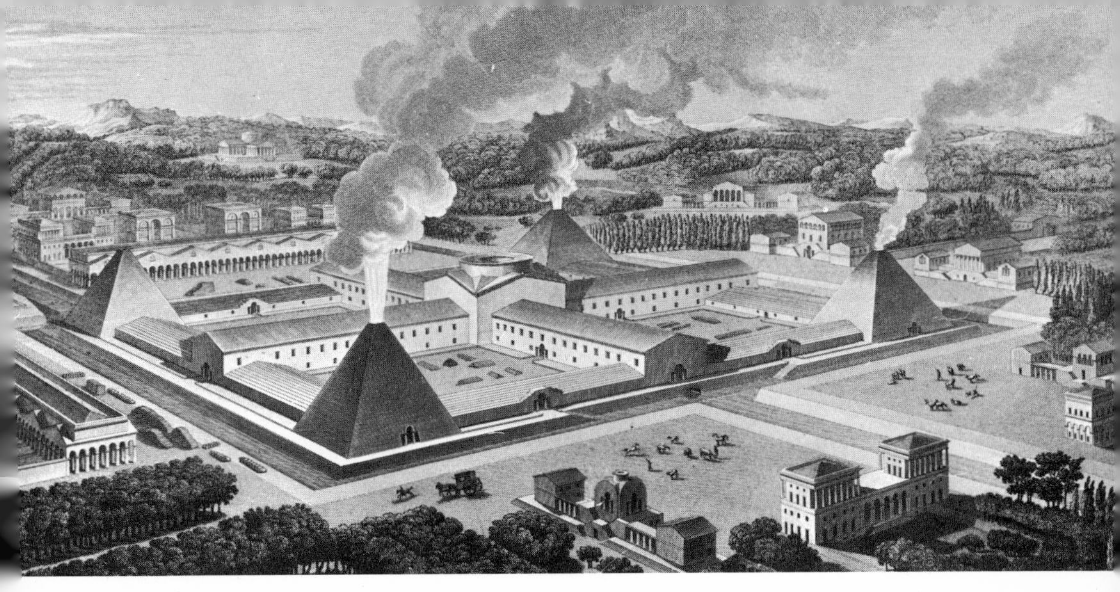

Fig. 84. "Forge à Canons" (project). Part of architect Claude Nicolas Ledoux's vast utopian design for a model industrial city for royal monopolies.

Fig. 85. Plan, Palace of the Phalanx, c. 1821, as published by the Utopian socialist, Briancourt.

Fig. 86. Bird's-eye view, New Harmony, Ind., c. 1825. Stedman Whitwell, architect for Robert Owen.

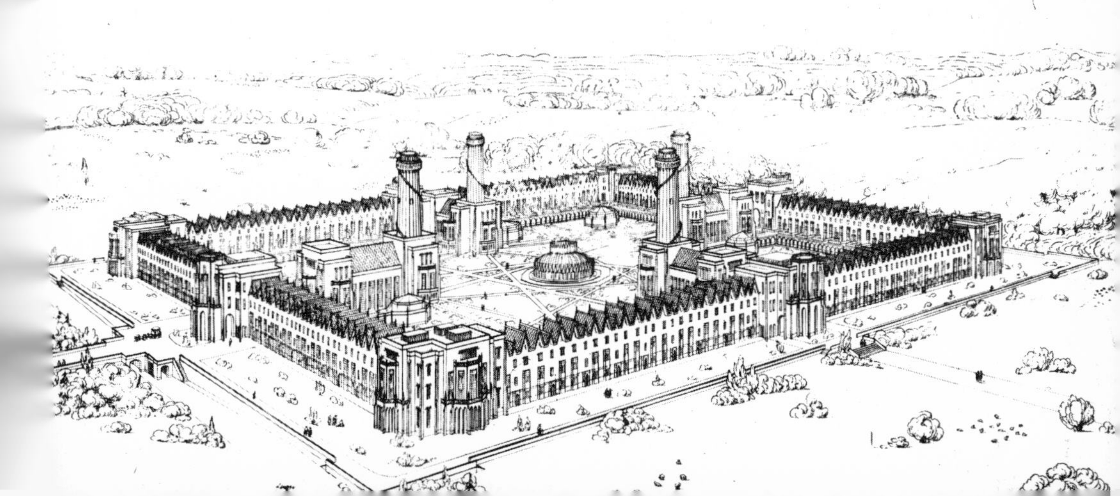

continued application of industrialism to human needs. They therefore urged its rejection — escape to the putatively happier Middle Ages. The Romantic poets and painters oscillated between the two extremes. Sometimes they were advanced in their social and political orientation (Blake, Delacroix); sometimes they were obscurantist and idiosyncratic (Allston, Poe); but almost without exception, they rejected the contemporary landscape, looking either toward the far future or the long ago, the far away or the inner reaches of the imagination. The striking thing about so much of the nineteenth century is that, though many of these men had something new or original to say, none of them had a new or original language in which to say it. They were merely united in rejecting the Classic idiom as being no longer viable. They thereby prepared the ground for the stylistic eclecticism which was to dominate the century.

Eclecticism may be defined as the conscious and deliberate adoption and use, by an artist of one period, of the mode of expression of another epoch not his own. In one sense, of course, the architecture of Western Europe and the New World had been revivalist since the opening of the fifteenth century, when the architects had adopted a visual language which derived from Greco-Roman antiquity. But this process had gone on for so many centuries that, for all its many dialects, the Classic had become the absolutely universal language of the West. Everyone, cultured and illiterate alike, employed some variant of it when he built. When Jefferson turned back to the Romans for architectural inspiration, he saw them not as exotic or remote but as ancestors of whom he was a direct and lineal descendant. The new eclecticism sought deliberately to rupture this continuity. But it was an act of formal rather than substantive significance; and before it had run its disastrous course, it was to leave not a stone of the past unturned in its search for other styles.

GOTHIC AND GOOD — THE GLIB ANALOGY

For many reasons, the Gothic Revival proved to be the most powerful and extensive stylistic movement of them all. It came to America from England, where its origins were more literary than artistic. The first interest in the Middle Ages appears in such gentlemanly antiquarians as Horace Walpole and Sir Walter Scott. Walpole remodeled his home, Strawberry Hill, in the Gothic manner in the 1750's; Scott placed many of his novels in Gothic times; and while neither house nor novel bore much relation to the harsh realities of medieval life, they set in motion a portentous train of events. Already at the opening of the nineteenth century, the movement had begun to acquire momentum in England. It was, at first, largely secular. It was put to private service in such country houses as the one which the millionaire eccentric William Beckford had the architect James Wyatt build for him at Fonthill Abbey. This preposterous pile, erected largely of wood and plaster, was a sort of three-dimensional stage set designed to give an *arriviste* a spurious baronial background. Finished after 1800, it collapsed in rubble and dust (in 1823) around the ears of another owner (Figs. 87, 88).

But the Gothic Revival soon began to take on nationalistic overtones in both England and France. In England, this patriotic identity was sharply etched by the controversy about the reconstruction of the old Houses of Parliament after their destruction by fire in 1834. In France, it was connected with the national program for restoration of such medieval monuments as Sainte Chapelle and Notre Dame in Paris and the walled city of Carcassonne. On both sides of the Channel, the Gothic Revival was thus to become an ideological arm of national patriotic ambitions.

Ultimately, of course, the movement was to take on overwhelmingly religious connotations. In one sense, this was simply associational since the Gothic had always reached its highest expression in religious buildings and, even in its secular examples, was always the idiom of theocracy. Moreover, in England at least, the Gothic Revival was directly connected with a religious revival in the established church called the Cambridge Movement. It was here, as an undergraduate, that the young Pugin was launched on his career as the leading exponent for the Gothic Revival. His arguments for the style, like those of Ruskin after him, were carried on at two different levels — one rational, the other moralistic. His defense of Gothic as a system of construction was apt and vivid: it was structurally efficient, constructionally honest, artistically coherent

Fig. 87. Strawberry Hill, Twickenham, London, begun 1748. Walpole was one of the first "Gothicists."

Fig. 88. Fonthill Abbey, Wiltshire, England, 1800. James Wyatt, architect. Badly built, it collapsed in 1823.

and unified. But his effort to prove that Gothic was the style of building uniquely suitable to Christian worship (while that of the Renaissance was "pagan") was simply grotesque. Yet, anti-historical as his thesis was — the Roman Church to which he was a convert had through most of its history been housed in Classic or proto-Classic buildings — it nevertheless received uncritical acceptance on both sides of the Atlantic. Protestant and Catholic alike accepted his typically Victorian analogy, as false as it was glib, between goodness and Gothic.

The resurrected Gothic received the final accolade of both secular and clerical approval when it was decided to reconstruct Westminster Abbey in that style in 1836. Actually, the decision seems to have appeared more or less inevitable to those men charged with the task. The old Abbey had been a very catalogue of English experience with the style from the days of Henry III to the time when Henry VIII added the Henry VII Chapel in 1509. It thus spanned the rise of the modern English nation. And though hardheaded commoners like the Utilitarian Jeremy Bentham might be on record for comfort and economy in the rebuilt Parliament, there seems to have been little feeling that Gothic was not the appropriate mode in which to achieve both.[20] In any event, as the most important building in the British empire, under construction from 1840 to 1860, it threw an irresistible prestige behind the style. It thus became the absolutely regnant idiom of polite society in England and — by rapid and easy extension — in the United States.

As a matter of fact, interest in the Gothic had already been apparent in this country at the turn of the century. The French *émigré* architect Maximilien Godefroy had built the little Gothic chapel of St. Mary in Baltimore in 1807. Benjamin Latrobe, for all his lucid classicism, had submitted a "Gothic" design for the new Catholic cathedral in the same city in 1806. Even Thomas Jefferson, in his old age, had toyed with a little Gothic folly for the family graveyard. But it was not until the late thirties that the style began to appear as a serious contender with the Greek — and only then in the East. Just as in England, the arguments in its favor are just as often pietistic as esthetic. Thus one American critic rejoiced in the discovery of —

a style of architecture which belongs peculiarly to Christianity . . . whose very ornaments remind one of the joys of life beyond the grave; whose lofty vaults and arches are crowded with the forms of prophets and martyrs and beatified spirits, and seem to resound with the choral hymns of angels and archangels . . . the architecture of Christianity, the sublime, the glorious Gothic.[21]

Such arguments were a Protestant American paraphrase of Pugin and anticipated, by a decade at least, Ruskin's immensely influential argument along the same line. How persuasive they were with the American upper classes became obvious during the forties, both in what they built and in what they wrote about the buildings. Gothic churches and villas became the rage; New York University was rehoused in Gothic; it was even being tried out on commercial and industrial projects. And the *cognoscenti* were becoming proud of their ability to discriminate between good Gothic and bad. Thus the Manhattan diarist, George Templeton Strong, found the First Presbyterian Church "an abortion . . . just such a travesty of a Gothic church as one might expect from a bankrupt Unitarian builder of meeting houses." Renwick's Grace Church showed an "unhappy straining after cheap magnificence"; but the same architect's St. Patrick's he found "very ambitious; scale very grand indeed — likely to be effective." He felt, however, the use of "cheap ornamentation in iron" would "surely rock [the church] to pieces by expansion and contraction of its incongruous materials." Upjohn's Trinity Church, on the other hand, struck him as "altogether the finest interior I ever saw, the only Gothic interior that ever seemed *natural* and genuine and not the work of yesterday."[22]

But, in the shadow of the Gothic, other antique styles were clamoring for attention. All of them were echoes, either literary or touristic, of American contacts with Europe, Asia, India, Egypt. Napoleon's conquest of the latter is reflected in Haviland's New York City Jail and in Strickland's First Presbyterian Church in Nashville. The mandatory visit to Florence produced a rash of "Tuscan" (or, as the architects Town and Davis called them, "Etruscan") villas. British conquest of the Taj Mahal is echoed in the "Persian" architecture of Longwood Plantation in Natchez, Mississippi. There were "Swiss" chalets and "Heidelberg" towers and they were all to con-

Fig. 89. The New Palace of Westminster, 1840–65. Sir Charles Barry, architect, A. W. Pugin, designer.

Fig. 90. Grace Church, New York, N. Y., 1843–1846. James L. Renwick, architect.

tribute to the unparalleled stylistic confusion of the last half of the century. But none of them approached the prestige of the Gothic because none of them was buttressed, like the Gothic, with a pietistic rationale.

Finally, there was yet another solvent of Classic hegemony in the Romantic alchemy: the Picturesque landscape. As the very term implies, this attitude was first given visual form by the painters: but, like most artistic movements of the period, it was literary in origin. Jean Jacques Rousseau, with his noble savage in an unspoiled natural landscape, had laid the basis for a new attitude toward Nature among Europeans. Instead of their traditional distrust of her as the habitat of hostile and malign forces, they began to think of her as benign, the source of spiritual as well as material nourishment. Goethe was the poet of this new view of the landscape, Constable and Corot its painters. And, taking literally the precepts of poet and painter that Nature unadorned was the paradigm of beauty, a succession of English gardeners, from Humphry Repton to Joseph Paxton were busily at work ripping up formal Renaissance gardens and replacing them with naturalistic designs.

The process was repeated on this side of the Atlantic, allowing for the usual time lag of three or four decades. Fenimore Cooper's novels gave Americans a local version of the noble savage in his unspoiled wilderness. The Hudson River school of painters provided the pictorial image of this new landscape and Dickens the perfect Victorian rationale, written when he first saw Niagara Falls in 1842:

> . . . then, when I felt how near to my Creator I was standing, the first effect, and the enduring one — instant and lasting — of that tremendous spectacle, was Peace . . . calm recollections of the Dead, great thoughts of Eternal Rest. . . . What voices spoke . . . what faces, faded from the earth . . . what Heavenly promise glistened in those angels' tears![23]

And the circle was closed with the books and gardens of Andrew Jackson Downing, ambitious and talented son of a Hudson River nurseryman, who made himself into our first professional landscape architect. Eloquent though not original, he very successfully paraphrased English theories of the Picturesque landscape for the benefit of the new American suburbanites. Though he himself died

Fig. 91. Residence of J. B. Choller, Watervliet, N. Y., c. 1848.

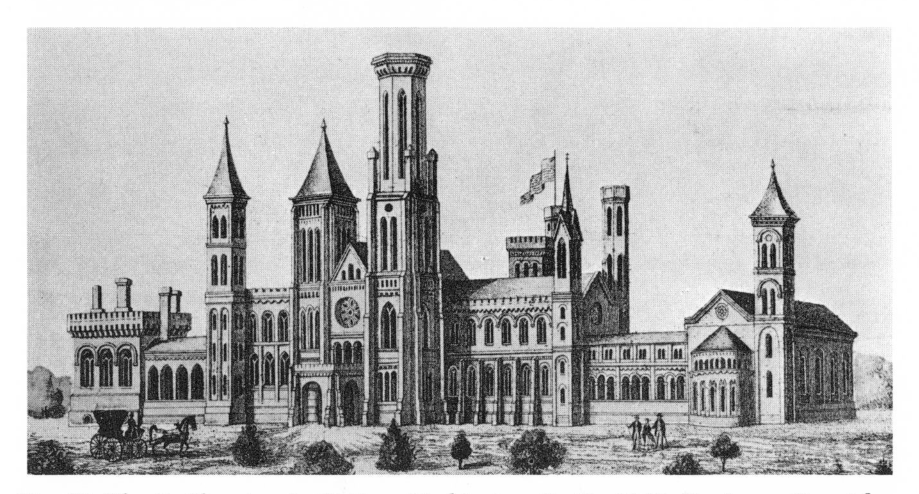

Fig. 92. The Smithsonian Institution, Washington, D. C., 1848–49. James Renwick, architect.

Fig. 93. Halls of Justice ("The Tombs"), New York, 1836–38. John Haviland, architect.

young, his work led directly to the creation of Central Park and formed the basis for most landscape theory during the latter half of the century. Thus eclecticism shattered the Classic image in landscape design just as it did in architecture proper.[24]

Insofar as Romanticism served to liquidate the formalism of a defunct Classicism, it served a historically necessary function. Whatever its motivations, its liberating effect upon planning was healthy: so too was its new emphasis upon the out-of-doors. As long as the basic plan types and structural systems remained unchanged, the use of historic systems of ornament remained a dignified occupation. Gentlemen could split hairs in comfort and safety. But, unfortunately for the case of the Romantics, merely to switch from one set of derivative esthetic standards to another equally artificial was no longer sufficient. For history had raised a new question: not *which* of these familiar styles was most suitable, but rather were *any* of them any longer useful?

Fig. 94. Wedding Cake House, Kennebunk, Me., remodelled c. 1850. A fantastic jig-saw wooden screen has been wrapped around a four-square brick house c. 1800. Note "Gothic" service wing, right.

4. 1840–1860

THE SCHISM

The motive force of Victorian life — that force of which all stylistic controversy was merely a surface manifestation — was the industrialization of the Western world. This process left not a concept, an institution or a building unaffected. Its impact upon architecture can be traced at many different levels: two of the most revealing are the morphological changes in the city and in the individual dwelling house. In a very real sense, these are but two aspects of the same phenomenon — the public and private aspects of man's abode. Improvement in the material conditions of life was one of the great goals of the century: freedom from pestilence, from fire, from cold and hunger. But these were collective goals, obtainable initially only in the city. (As we shall later see, the urbanization of the countryside awaited electric power and the internal combustion engine.) With the rapid growth of American cities on the one hand; and the increasingly disastrous fires and epidemics which swept them on the other; it is not surprising that the public and private amenities listed on page 108 were perfected during the nineteenth century.

Industrialization and urbanization radically altered the social and economic landscape in which architecture developed; and it simultaneously altered both the function of buildings and the structure, the very materials out of which they were fabricated. This is particularly clear in the American house since it was the nexus of changing patterns of family life. Changing theories of health, hygiene, and

THE TWO FACES OF URBANIZATION

PUBLIC AMENITIES (in the street)	PRIVATE AMENITIES (in the building)
Piped-in water	Running water
Sewers { septic storm	Waste disposal
	Kitchen sink
	Indoor laundry tubs
	Complete bath { hot and cold water flush toilet tub and shower
Paved { sidewalks streets	Easier housekeeping: reduced { sweeping dusting mopping
Street cleaning	
Snow plowing	
Garbage removal	
Trash removal	
Stray animals removed	
Street lighting	Central lighting
Fuel delivery { wood coal kerosene gas electricity	Central heating
Food delivery { ice milk meats and groceries	Domestic refrigerator
Mass transport { jitneys horse cars trolleys trains	
Policing	
Firefighting	
Public { drinking fountains toilets baths	

sanitation; of child and family care; of cookery, diet, and nutrition; even the woman's struggle for equality, from the right of suffrage to the right to labor-saving devices in the home; all of these had important effects upon the house. Qualitatively, the shift was from the rural to the urban, from handicraft to industrial manufacture, from a center of production to a center of consumption. It was a long and complex process, not to be actually completed until the middle of the present century. But by 1840, it was already well along.[1]

Of course, the overwhelming majority of American families were still farm families. And the farm family of that day was a self-sufficient economic unit to an extent difficult to imagine today. The farmhouse reflected this fact. Food was not merely prepared for the table there — it was manufactured there. The kitchen was a food factory, family run. It was the processing center of a network of facilities: icehouse, springhouse, well house, milk house, smokehouse, washhouse, root cellar, and woodpile. Further out lay the orbit of vegetable garden, orchard, cow barn, pigpen, chicken run, corncrib, haystack, and barns. Beyond this lay the farm itself.

It was not, by modern standards, a very efficient industry. The whole family worked, sunup to sundown, to keep it going. Without refrigeration, meat could only be preserved by smoking or brine. With only the springhouse for cooling, milk could be preserved only as butter or cheese. Home canning of fruits and vegetables was not possible until the appearance of the glass Mason jar in 1858: prior to that, the only method of preserving fruits was by drying or in jams and jellies using sugar too expensive for most housewives. There was no means of extending the short periods when green vegetables were available, though root cellars kept potatoes and turnips fairly well. All lard was home rendered, all bread home baked. Thus, though a prosperous farm might provide a calorically adequate diet, it was not apt to be a very well balanced one, particularly in winter.

The goods and services produced on the farm extended far beyond the family's food: the manufacture of cloth and clothing, quilts and bedding; of shoes and furniture; blacksmith work for the farm; even the coffins for the family dead. And in addition to cooking and washing for the family, the housewife's chores included nursing the very young, the very old and the sick. The texts on household management by Mrs. Childs, Miss Leslie, and Miss Beecher offer a poignant catalogue of the wife's responsibility.[2] Indeed, whole sections of these books deal with physiology and hygiene, with household chemistry and domestic pharmacopoeia, which today are the province of the doctor, the pharmacist, and the hospital. They include instructions (much of it tragically mistaken) on how to treat rabies, cancer, snakebite and fever; on how to dye cloth and clean gloves, make soap and candles; on how to decorate the parlor and ornament the dooryard.

But by the mid-century, the house in the big cities had ceased to be a self-sufficient unit. Though many a housewife might still manage to keep some chickens, or a pig for the family slops, or a milk cow, the urban house was already nourished by a network of services for its food, its fuel and water. A critical problem for the growing urban populations was the provision of a safe supply of fresh meat, fish, milk, and green stuffs. Both pollution and putrescence were involved, though the connection between them and illness was only obscurely understood. Protective measures were largely empirical, as indeed they had to be before the basic work of Pasteur in microbiology; considering this, they were on the whole surprisingly effective. Nevertheless, the record is full of tragicomic episodes. Thus, at the height of the cholera epidemic of 1832, the City Council in New York passed an ordinance forbidding the sale of *any* fresh fruits and vegetables; and the Chicago papers, in 1849, reprimanded the local lawmakers for not doing the same thing.

One important step in breaking this cycle of putrefaction and disease was the rising use of ice for refrigeration. Though ice had been used for millennia to chill wines and freeze sherbets, its use in food preservation seems to have been almost an American invention. European visitors were much struck by it. Adam Fergusson, an English traveler, reported in amazement that, in the hot summer of 1831, his hosts in Washington "contrived to preserve untainted their fish and butcher meats" in a household refrigerator. Count Arese found the icing of milk commonplace throughout the Midwest in the same decade. The icehouse had been fairly common on prosperous farms but prefabricated household refrigerators were described by the New York *Mirror* as "an article of necessity" in July, 1838. And Miss Leslie's *House Book* of 1840 called them "conveniences no family should be without." They were being manufactured in quantity in Boston and sold all over the East in the same year.

Parallel with the spread of the refrigerator came the harvesting and distribution of natural ice to urban homes. It was for sale in Cincinnati in 1830, in Philadelphia in 1833, and generally throughout northern cities by the 1840's. Boston ice was being imported into Charleston, Savannah, and New Orleans in 1817. The harvesting of ice became a great industry in Boston, where ice harvesters began shipping it to the Caribbean, then to London and finally, via the Clipper ships, to the seaports of China. At its peak, in 1870, this trade amounted to some 15,000,000 pounds.[3]

But industrialism was simultaneously setting into motion another and much more profound revolution — the mass production, processing and distribution of food itself. This process can be traced with fascinating clarity in Chicago. Cyrus McCormick moved his reaper plant to that city in 1847. In 1849, he turned out 1500 machines. In that year also the first railroad entered town. A decade later, McCormick production was up to 5000; and a dense network of railroads linked Chicago to the prairies. Mass-produced grain made possible mass-produced livestock; that in turn led to meat-processing centers. Initially a cold weather operation, meat-packing soon had to be placed on a year-round basis. Hence refrigeration and canning techniques came in for accelerated development. Refrigerated warehouses using natural ice were already commonplace by 1850; and as early as 1857 the first refrigerated car appeared. In 1843, the railroads brought five million quarts of iced milk into New York. In 1866, they brought the first out-of-season vegetables from the Gulf Coast to Chicago. In 1870, refrigerator cars brought the first vegetables, grapes, and salmon to Chicago from California. Meanwhile, for the far South which had to import all its ice, mechanical ice manufactories were developed. New Orleans already had one in operation by 1850; they were commonplace after the Civil War, though ice remained a luxury product until the turn of the century.

Fig. 95. Opening of the Croton Waters, 1842. A series of disastrous plagues had dramatized importance of pure drinking water to New Yorkers and need for this, the first public water system.

Fig. 96. Plumber's advertisement, c. 1842, shows uses of Croton Waters: fire-fighting, public fountains, horse-watering, drinking fountains and complete private bath including recessed tub, flush toilet and lavatory.

With this exploitation of refrigeration, along with similar developments in canning technology, industry had thus intervened at two different points — the home and the factory — to protect the household from putrescence.

THE SANITATION SYNDROME SOLVED

Public water supply and sewerage systems were also being developed in the cities. Adequate water supply was urgent on three distinct counts: for drinking, to replace contaminated wells and cisterns; for bathing, which was no longer regarded as hazardous; and for firefighting. The construction of a centralized water system, supplying nominally potable water under pressure throughout the city, was already quite practicable technically. (The chemical purification of water was to come much later.) Philadelphia had completed a system of reservoirs, steam pumping plant and wooden mains before 1820. When he died in New Orleans in 1819 (ironically, of one of the very endemic diseases which modern sanitation would eliminate), Latrobe had already begun a similar water system there. Cincinnati had such a system in operation by 1838 and New York had brought Croton waters downtown by 1842. But, though all the components of the system might be theoretically available, it would still be decades before they became universal. The costs were extremely high, largely because there was no mass-produced supply of steel conduits, mains, valves and piping. The complementary side of the sanitary problem — sewers for the disposal of body wastes and storm waters — was subject to the same limitations. Here, too, the need outpaced the technical resources of the country.

The development of central water supply and sewerage systems, securely sealed against cross-pollution or outside contamination, was closely connected with another set of urban problems — i.e., the cleanliness of streets and sidewalks. Here, by all accounts, conditions were appalling, even in the centers of the largest cities. James Fenimore Cooper wrote "his dearest wife Sue" that walking down Broadway "was walking in the mud, and soft mud, too . . . the town is supremely dirty."[4] Walt Whitman complained of being splashed by mud from passing carriages on the same fashionable thoroughfare. And scavenging pigs, goats and geese led one Manhattan wag to pen this jingle for the press:

> A question 'tis, and mooted strong
> Between the citizens and swine,
> To which the streets do most belong,
> Of this most glorious city.[5]

In a more serious vein, G. T. Strong wrote in his diary that the old Potters' Field at Fourth Avenue and Fiftieth Street in Manhattan was "a disgrace and a scandal . . . dead paupers gnawed by the gaunt swine that are co-tenants with the Hibernian humanity of the adjoining shanties. This is within a hundred yards of a dense population."[6]

In the absence of municipal collection and disposal systems, garbage and trash accumulated in streets, alleys, and yards. Horses added their feces to that of the scavenging livestock. Even where street sweepers were available, they would often be powerless in the face of streets that were alternately deep dust, deep mud, or deep snow. The obvious solution to this dilemma was a continuous pavement — hard, smooth, and permanent: yet such surfacings, in the form of concrete or asphalt pavements, were still half a century or more away. Until then, brick or cobblestone paving would have to do, inadequate and expensive as it was.

With the appearance of water and sewer systems, the development of the bathroom was a natural corollary. Above all, they made possible the water closet, the first safe and esthetically satisfactory method of disposing of body wastes. Moreover, the importance to health of bodily cleanliness was becoming generally understood, so that the tub and shower bath were no longer under interdiction. As a result, the modern bathroom was appearing in the big metropolitan hotels and the city houses of the rich. Kitchen sink and laundry tub were also becoming familiar features of the urban house. However, such amenities were not common until decades after the Civil War and not really standard equipment until the present century. The reasons for this discrepancy between potential and reality were again largely practical: plumbing would remain a luxury as long as it was a largely handicraft affair.

As early as 1807 the citizens of Edinburgh had been startled half out of their wits to see William Murdock, local scientist and inventor, driving around at night in a steam-powered buggy whose headlights were fed by flaming pig bladders filled with coal gas. This was the same Murdock who had piped gas out to a lamp in front of his house, thus initiating modern street lighting. It was he, too, who spent years in perfecting a practical method for producing gas from coal and installed the first industrial lighting system (in an English foundry in 1803). His work had early repercussions. By 1812 London had its first gas-lighting company; and three years later Philadelphia — always alert to municipal improvement — was trying to launch a similar undertaking. Some trouble developed, however, and New York and Baltimore were both to get gas lighting before the City of Brotherly Love finally opened its plant in 1835.

The advantages of gas lighting — especially in industrial and commercial buildings — were so obvious that it soon became a standard. By 1860 every large city in the country depended upon gas for its principal source of artificial light. There was a steady improvement in burners, mantles, shades, etc., which constantly increased the efficiency of gas lighting. Most of these improvements originated abroad, but were almost immediately adopted in this country. The result was that, by the time of the Civil War, an entirely new concept had appeared in building design: that of a fixed, semi-automatic lighting system which *freed the building from its historic dependence upon natural daylight.* This was of enormous significance to a society whose buildings had hitherto lain idle at least half the time. It opened up the possibility of buildings which could be used on a twenty-four-hour basis, of industrial and commercial processes which could run uninterruptedly around the clock. It was, in other words, the first step toward a complete synthetic luminous environment.

Gas began to light the textile factories of New England; and factory owners were able to extend the working day to ten, twelve, fourteen hours a day. Gas began to light the theaters of New York; and the "close-up" appeared for the first time in dramaturgy. Gas began to light the city streets; and a whole medieval world, in which all streets at night were more dangerous than a jungle path, disappeared from man's consciousness.

But the production and distribution of gas was a complex and costly operation. Its main use was in commerce, industry, and transportation. Only the rich could afford it in their houses. It did not begin to light the homes of the common people until the last part of the century, when electricity had already rendered it obsolete. Even more important, gas lighting was always purely urban. Two-thirds of the American people had no access to it at all. Instead, they used the oil lamp. By 1830 the lamp had assumed approximately its present form — cloth wick, glass chimney, and shade. The fuel was whale oil. The huge and daring whaling fleet which roamed the seven seas in American-built ships was based on this lowly lamp. The fleet rose with it and fell only after 1859, when Pennsylvania's Drake drilled the first petroleum well in history and the manufacture of kerosene began.

Like many commonplaces of modern American life, central heating is neither American nor modern in its origin. The idea of heating buildings by pipes filled with hot water or steam must have appeared shortly after James Watt perfected the steam boiler. At any rate, by 1836 London had enough experience to enable Thomas Tredgold to publish a book on the subject, which gave substantially correct methods for computing "loss of heat radiated from buildings and corresponding methods for computing the necessary size of radiating surfaces." Tredgold used spherical boilers of cast iron and cylindrical boilers of wrought iron. Although equipped with safety valves, these resembled nothing so much as huge and treacherous teakettles set in brick. In England, these were put to work heating the glasshouses of royalty and little else.

Until well after the middle of the century, central-heating systems were forced to remain on a custom-built basis, with the heating engineer acting as his own designer and manufacturer of pipes, couplings, radiators, and boilers. Nevertheless, Americans displayed a keen interest in the relatively great efficiency of steam and hot-water heating, and it was they who eventually put it on a mass-production basis. Here again they could borrow from the factory, where the use of steam power forced the development of boilers, piping, etc. New York opened its new jail in 1840 — "a dismal-fronted pile of bastard Egyptian, like an enchanter's palace in a melodrama," said Dickens (Fig. 93). It was immediately and appropriately called the Tombs; but it had central heating, and scarcely three months had passed before a criminal had hanged himself from the pipes in his cell. Steam-heating systems had begun to appear here and there, but by and large heating could not appear on any wide scale until industry could provide the boilers, piping, radiators, etc., and it was not until the mid-Victorian period that this would be possible.

Control of the thermal and atmospheric environment was simultaneously proceeding along another line, however. Franklin's stove had given birth to a prodigious offspring. The cast-iron cooking stove was rapidly becoming basic equipment for every kitchen, the cast-iron heater for every parlor. Dickens found it "common to all American interiors . . . the eternal, accursed, suffocating, red-hot demon of a stove whose breath would blight the purest air under heaven." For larger buildings, the principle of convected heat had already shifted from stove to furnace; and the concept of air conditioning was not far away. As early as 1840, Robert Mills, architect for the Capitol and one of the great Jeffersonian school which combined architect with mechanic, had patented a system for cooling the air in summer, warming it in winter. Such a system was actually installed in the United States Capitol a few years later but there is no record of it in operation. In the 1840's a Florida physician, Dr. John Gorrie of Apalachicola, was treating malarial patients. In an effort to save them both from the southern heat and their own body's fever, he conceived the idea of cooling the hospital rooms. Using the now familiar compressor, he rigged up a working model of what was probably the first mechanical cooling unit in history. He demonstrated this model to an unfriendly press and his experiment came to nothing. It was to be the meat-packing industry, in the post–Civil War years, which would furnish the principal incentive to artificial ice and refrigeration.

Few buildings had central heating, and a long time would elapse before it was a standard aspect of our building. Yet two treatises had appeared in 1844 which showed that the fundamental principles of heating buildings by hot water and hot air were well understood: Charles Hood's *Warming Buildings* and D. B. Reid's *Theory and*

Practice of Moving Air.[7] As in so many other instances, such works gathered dust on library shelves until the everyday world caught up with them.

DOMESTICATED UTOPIAS

Catharine Beecher, the most astute student of the American family throughout this period, published a series of house designs in her book of 1869. Stylistically, these houses are designed in a kind of *lumpen* Gothic which — good Victorian that she is — she claims "might properly be called . . . Christian." Be that as it may, her floor plans reflect quite accurately the new relationship of family to society, house to urban environment. Two qualities stamp them as essentially modern: the way in which all her enclosed volumes are designed to facilitate specific housekeeping tasks; and the masterly way in which she exploits the new urban services and absorbs them into the very fabric of her plans.

Her houses are firmly visualized as machines for family life or, as she puts it, "contrived for the express purpose of enabling every member of the family to labor with the hands for the common good, and [that] by modes at once healthful, economical and tasteful."[8]

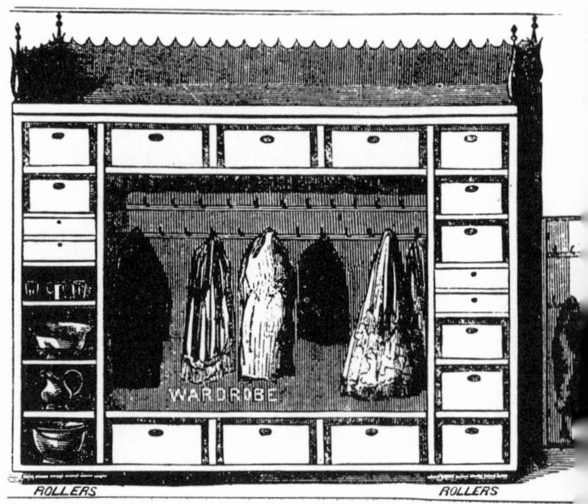

Fig. 97. Kitchen work center (left) and movable storage wall (right) show Catharine Beecher's grasp of need for organizing household tasks along lines of rationalized work in industry.

Fig. 98. Front elevation, model house, 1869. One of many designs published by Catharine Beecher to demonstrate principles of "domestic economy," i.e., lower middle class suburban servantless housekeeping.

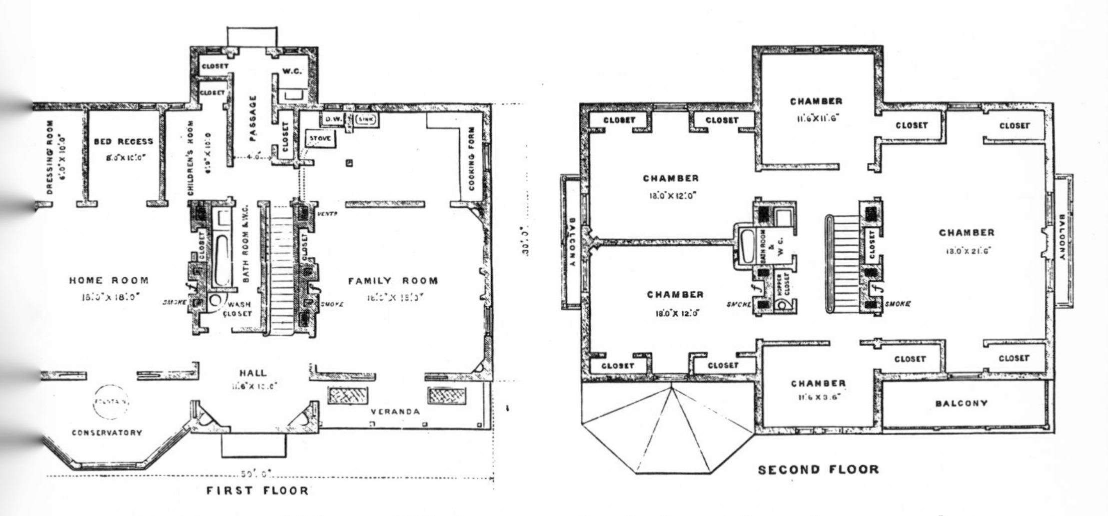

Fig. 99. Plans, model house, 1869, demonstrate her fundamentally modern approach to house planning, employing functional theories of classified storage, modern cookery, central heating and compartmentalized baths.

No longer are there generalized or anonymous spaces: from top to bottom, each area has been carefully organized to serve a specific function. Even her nomenclature is precise — "family room," "home room," "children's room," "bed recess," "flour bin," "cooking form," etc. Classified storage space is understood to be the first requisite of efficient housekeeping. In her kitchens, this leads to cabinets of astonishing modernity whose shelving, cupboards, bins, and work surfaces fully anticipate today's kitchen (Fig. 97). Her "home room" on the first floor is a multipurpose area designed for quick convertibility to meet a variety of domestic emergencies: an overnight guest, a sick child or aged relative needing nursing, a ladies' sewing party. The big dressing room and bed closet make such conversions easy. Then, for the young baby who was a recurrent feature of family life in the days before contraceptives, Miss Beecher provides a small room just off the kitchen. This reduces time and effort of periodic nursing, diaper changing, etc.

Her houses make a wide and intelligent use of the newly available sanitary facilities. This one not only has two bathrooms, one on each floor, but both are compartmentalized to permit simultaneous use of various fixtures. In addition, there is a separate W. C. off the kitchen, presumably for kitchen or yard help. The kitchen has a sink and the basement has a complete laundry. The modern cast-iron cooking stove is served by a dumbwaiter to bring fuel up from the basement and take ashes down. In a compartment of its own, it can be closed off when not in use.

This particular house still relies upon fireplaces for heating; but other designs in the same book employ stoves, furnaces, and fireplaces in different combinations. She advocates gas for lighting because it eliminates all the tedious wick-trimming and chimney-polishing connected with oil lamps. It is obvious from cupboard space in kitchen and basement that this is an urban family, dependent for its food upon the markets. There are no provisions for the canned foods, root vegetables, and smoked meat that would be invariably found in the farmhouse. Some of her houses show built-in refrigerators.

Here, in short, in the very year of his birth, is the prototype of that urbanized house which Frank Lloyd Wright would raise to the level of a work of art. For all its graceless and inept details, it is the direct progenitor of the modern American house.

Industrial technology had also modified the physical structure of Miss Beecher's houses. They were built of wood, as the overwhelming majority of domestic architecture always had been. But they employed the new system of construction known as the "balloon frame." Unlike earlier framing methods, this used light dimension lumber nailed together, as against the mortise-and-tenon and pegged connections of earlier work. In a masterpiece of historical sleuthing, the late Walker Field traced the actual invention of this system to Augustine Taylor, a Chicago builder who first employed it in a little church in 1839.[9] In retrospect, however, the invention of the balloon frame seems as inevitable as that of the steel frame in the skyscraper, and for almost identical reasons. Skeletal construction was more efficient and its fabrication easier to rationalize than any other sort. The balloon frame merely awaited two technological advances: the mass-production and distribution of dimensioned lumber and machine-made nails. According to Horace Greeley's study, *Great Industries of the U.S.*, nails had cost more than screws in the seventeenth century; but by 1828, the cost was down to 8¢ per lb. and in 1842, to 3¢. Dimensioned lumber and cheap nails made possible a whole new order of speed and economy in wood framing. "A man and a boy can now attain the same results, with ease," said one writer, "that twenty men could on an old fashioned frame."

Industry had demanded two qualities of mill and factory from the first — greater span and greater strength. These demands flowed from the very nature of mechanized mass production and, with the general use of steam power by the 1830's, became imperative. For the economical use of steam implied not only that all stages of a given process be brought indoors, but more and more that they be concentrated under one roof. This meant greater spans, unobstructed floor area. Steam power also brought enormous increase in the size, power, and weight of machinery. This meant stronger floors and walls.

The traditional structural system soon proved inadequate. The limitations of the simple wood beam were such that experimentation with a composite element was inevitable. Hence the truss. Although implicit in the pitched roofs of the Pilgrims, the truss had only been employed in the larger public buildings, churches, and mills of the pre-Revolutionary period. The competition drawings for the United States Capitol had shown some excellent early trusses. It was now further developed as an independent structural element in the factory. But it was the bridge builder and not the architect who first saw its potentials. The railroads had begun their march across the continent; when they came to a river, they had to cross it — rapidly, cheaply, and safely. Their problem, like that of the factory, was greater span; like the factory, they could not afford the laborious masonry arch. Instead, they turned to the truss, which, under rapidly accumulating experience, ceased to be the intuitive creation of some pragmatic builder and became a formula in engineering. These early trusses were of wood.

The column was subject to no such refinement — its cross section must increase in direct ratio to its increased load. The limits of wood, stone, and brick had already been reached. What was urgently required was not a new form but a new material whose strength in compression was much greater. The answer came with iron — first *cast,* then *wrought,* and finally processed into steel. The cast-iron column was in use in England to replace wooden posts as early as 1780. Its appearance in this country seems to have been delayed by lack of adequate foundries, not ignorance. By 1840, cast-iron columns were appearing in New England factories and New Orleans business blocks. But already in 1830, Haviland had given a Pottsville, Pennsylvania, bank building an entire façade of cast iron; and the next two decades saw a wide development in such uses of this material.

If advance in the building field was made possible by the factory, it was by no means confined to it. The galvanic effect of industrialism was leaving its impress on all phases of life. In the rapidly growing cities all sorts of new building types were appearing to satisfy the qualitatively new needs of the population. Typical of these was the great metropolitan hotel, a type which appeared in full flower with Boston's Tremont House (Fig. 68). Although ex-

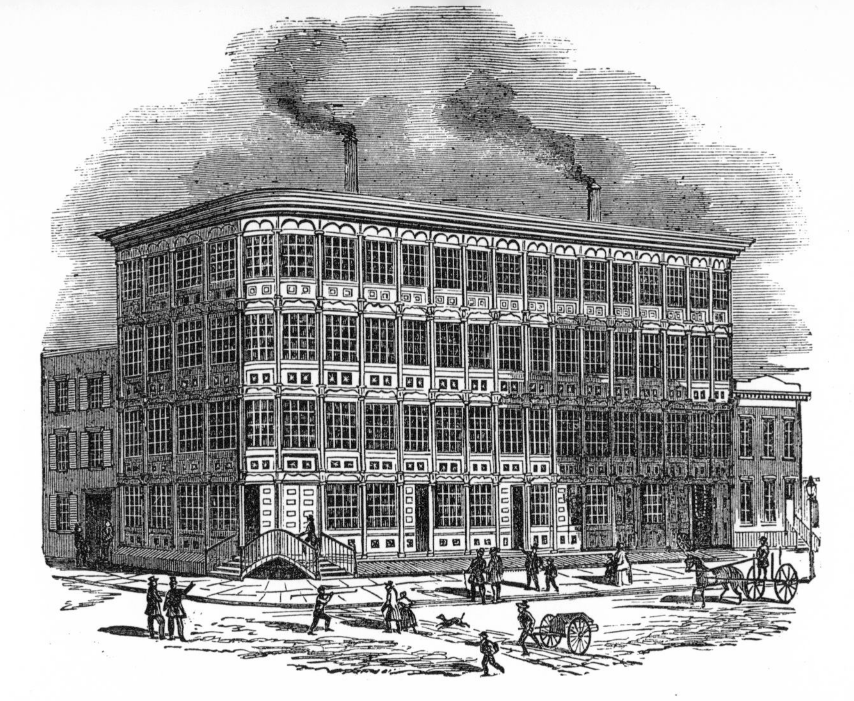

Fig. 100. Cast Iron Building, New York, N. Y., 1851. Designed, fabricated and erected by James Bogardus, this building had prefabricated frame and outside walls of cast- and wrought-iron members.

Fig. 101. Harper & Bros. Building, New York, N. Y., 1854. John B. Corlies, architect; James Bogardus designed and fabricated the metal structure to replace an earlier building destroyed by fire.

ternally this building preserved the decorous Greek proportions of its contemporaries, it was in fact the full-blown prototype of the modern Statler — complete with standardized obsequiousness, elaborate bathrooms, acres of gilded public space. Short-tempered Dickens, staying there in 1842 as befitted a famous visitor, was forced to admit that the Tremont House was "a very excellent one. It has more galleries, colonnades, piazzas and passages than I can remember, or the reader would believe." So immediate was the response of the traveling public (which was both rich and thick in these seaport cities) that the architect Isaiah Rogers found himself with a national reputation. In the 1830's, Mr. Rogers moved on to New York to build the even more lavish Astor House.

In fact, wherever one looked in the last years before the Civil War, one saw the reproductive powers of industrialism straining against the limits imposed by Classical architectural concepts and handicraft building techniques. It had already been demonstrated that it — and it alone — could reconstruct the fabric of American building. But actually to accomplish this task, science and technology no less than men themselves had to be liberated. Industrial production could not advance in a nation half slave and half free.

Clearly, then, our growing technology found the restrictions of traditional architectural conventions intolerable. In many fields — manufacturing, commerce, railroads, public works — they were already being discarded. On the one hand, new demands for greater height, increased span, concentrated strength to meet concentrated load, were making obsolete the old theories of structural design. On the other, many of the new building types had little, if any, ideological function to perform. Moreover, since efficiency was the paramount consideration, esthetic standards (as factors with objective, ideological importance) fell more and more into the discard under the pressure of pragmatic mill and shop owners. The purpose of a New England textile mill in 1840 was to manufacture cloth, not to sell the public on the importance of democracy (Fig. 102). *That* function was delegated to the public buildings which were springing up across the land.

To cope with these problems, which covered the whole of tech-

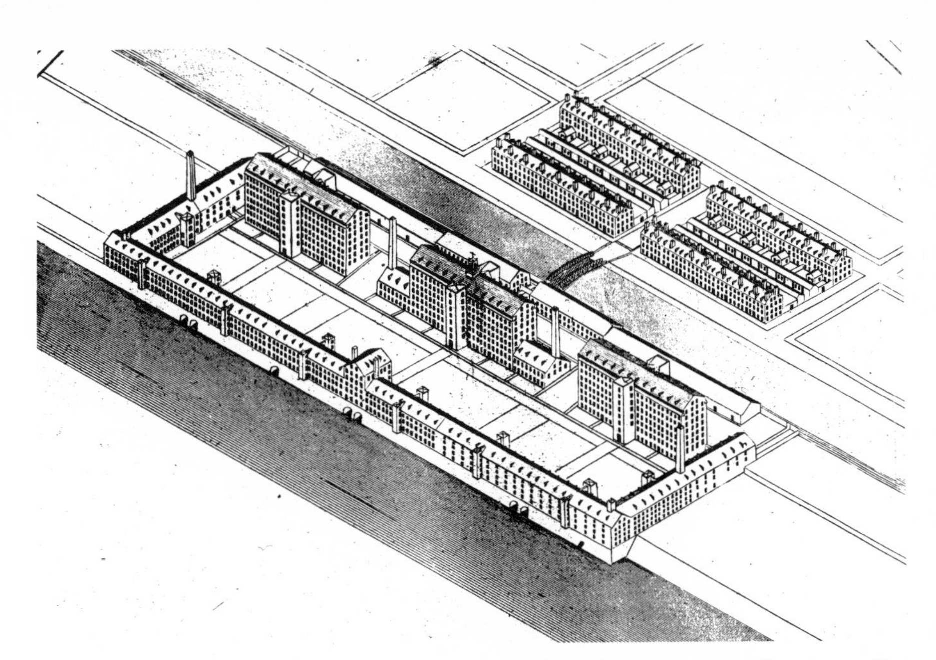

Fig. 102. Bay State Mills, Lawrence, Mass., mid-1840's. One of the first integrally-planned industrial complexes of mill, offices and model low-rent housing, including boarding houses for single women.

Fig. 103. Mississippi River Water Front, St. Louis, Mo., c. 1855. At the shear line between industrial north and agrarian south, this great commercial center also served as outfitting center for settlers headed for the far west. Ascendant technology is evident in steamboats and cast-iron buildings alike.

nology and were only partially those of building, society was creating a new professional — the engineer.[10] As the name implied, he was the lineal descendant of the "mechanick" of Jefferson's day, but with the theoretical training essential to the mastery of much more complex processes, machines, and materials. As contrasted with the architect, the engineer's assignment from the industrialist was the simpler: to build quickly, cheaply, and efficiently, and to hell with the looks. By and large, the early engineer was not a "cultured" man. His origins were the laboratory and the mill, the shop and the bridge gang. He was neither trained nor paid to explore the many fine shades of the esthetic problem over which the architects argued so learnedly. And as a system of formal education in engineering began to evolve out of the trades and mechanics' institutes of the early part of the century, his neglect of the "artistic" aspects of building design began to solidify into programmatic contempt. Indeed, this tendency went so far that, by the end of the nineteenth century, many of the specialized engineering and technical curricula were to be not merely anti-art but boorishly anti-cultural as well.

There was no conscious demand for "beauty" in the engineer's work. If, then, the American people began to see beauty in his designs (as they almost immediately did) it could only be because of his objective relationship to the productive processes themselves. They read into his work the fruitful implications of machine production. If it saved work, it could not be wholly ugly; and if it worked well, it would sooner or later be beautiful. The Americans understood (even if subconsciously) the new idiom which the engineer (also subconsciously) was beginning to develop. Form must spring not from preconceptions but from the limitations of actual material and process; form must be conditioned by the need for efficient performance in the finished product.

The appearance of the engineer as a professional with an independent status marks a decisive point in the history of American building — that point beyond which society's needs for control of its environment could no longer be filled by master builder or gentleman architect. By the same token, it marks the opening of that chasm between esthetic and technical standards which has characterized American building ever since.

JOHN RUSKIN, ROMANTIC TORY

As the Civil War approached, the entire structure of Classic tradition was under assault from many sides at the level of both fact and theory. By the Romanticists, under the Gothic banners of Sir Walter Scott, Pugin, Renwick, and Upjohn; by the early engineers like the elder Roebling, Eiffel, and technicians like Bogardus. But the most penetrating critiques were to come from two men, neither of whom was an architect and only one of whom was American: John Ruskin and Horatio Greenough.

Ruskin's impact upon American culture, especially our art and architecture, was immense. Indeed, as Henry-Russell Hitchcock has pointed out, his works received both earlier and greater recognition in this country than in England. And long after the initial impulse had ebbed, his influence was still to be traced in the Morris arts-in-industry movement, in Elbert Hubbard and his Roycrofters, in the tooled-leather-and-burnt-wood circles of the fin de siècle, in a hundred-odd byways and dead ends of upper-class neuroses and frustrations. The cumulative, objective effect of Ruskin's work was disastrous. This was all the more the case because he wrote for the layman. He was one of the earliest critics to direct his words, not at the closed circle of critics and intelligentsia, but at the middle-class audience which books, magazines, and rapid communication had created. With the possible exception of Pugin, there is perhaps no single theoretician in architectural history who has done more to deflect it into unproductive detours and cul-de-sacs.

Yet the intellect which Ruskin brought to bear upon the subject was deeper, more perceptive and subtle, than any which immediately preceded and most of those which followed it.[11]

Despite all his tedious deviations and asides, Ruskin explored the esthetic dimensions of building with remarkable consistency and thoroughness. For a lay audience, he defined his terms with unprecedented sharpness. He took his readers to the heart of the matter when he said "thus no one would call the laws architectural which determine the height of a breastwork or the position of a bastion. But if to the stone facing of that bastion be added an unnecessary feature, as a cable moulding, *that* is Architecture."[12] Ruskin meant, of course, that ornament was only *structurally* unnecessary. Architecture had other and "higher" functions to perform. The garlands and grapes, gargoyles and goddesses — why were they found only on the palaces, churches, capitols, and city halls? Why were they not equally distributed on city slum and country cottage? Why the vast and demonstrable difference between the anonymous body of popular building and the signed works of the architect? What was architecture, in short, but a specialized means of communication, "above and beyond the common use"; a set of symbols, idioms, and concepts with which the rich and mighty spoke to themselves and the world? The architecture of the nation, Ruskin said, "*is the exponent of its social and political virtues.*"

> It is very necessary, in the outset of all inquiry, to distinguish carefully between Architecture and Building. . . . Let us, therefore, at once confine the name [architecture] to that art which . . . impresses on the building certain characters venerable or beautiful, but otherwise unnecessary. . . . It may not be always easy to draw the line sharply and simply; because there are few buildings which have not some pretense of colour or being architectural; neither can there be any Architecture which is not based on building; but it is . . . very necessary . . . to understand fully that Architecture concerns itself only with those characters of any edifice which are above and beyond its common use.[13]

Here Ruskin showed the uncanny precision which often characterized his criticism. He had discovered one of the basic laws of motion in the history of man's building. But he had discovered this law at a critical stage of its history. Never before had the ideology imposed upon a nation's architecture by ruling-class necessity so flatly violated the contemporary building technology. The cardinals with their cathedrals, the emperors with their circuses, the tyrants with their tombs — whatever their motives, they had not been guilty of *structural* perversion. Stone angels never weakened the Gothic vault; caryatids were satisfactory columns as well as lovely maidens; since Assyrian walls were going to be surfaced in glazed tile anyway, it was a simple matter to bake in the azure lions and scarlet tigers.

The world in which Ruskin lived was quite otherwise. For several centuries the living fabric of western European building had been impressed into the Classic mold. For a long while this had been a reasonably successful expedient; proof of that lay in the fact that the Classic idiom had penetrated down into popular building, far below the reach of scholar or architect. But by 1840 the development of building technology had made a hollow (to Ruskin, obscene) mockery of the Classic style. The architecture around him was a living falsehood, guilty of both structural and moral perversion. Even the healthy body of popular building was being infected. Radical measures were called for if anything was to be saved.

Ruskin's art and architectural criticism was an organic part of the main body of his work. And however biased his view or superficial his penetration may often have been, Ruskin was a student of society itself. It was Victorian society, not just its esthetic standards, which appalled him. He represented aristocratic discontent with an industrialism which was, in England, already fairly complete. He registered the horror of cultured Englishmen at the appalling discrepancy between the promise of the machine and the actual squalor, suffering, and ugliness of nineteenth-century capitalism. It was a "carnivorous" system, based upon an "ossifiant theory of progress." "Our cities are a wilderness of spinning wheels instead of palaces; yet the people have no clothes. We have blackened every leaf of English greenwood with ashes, and the people die of cold; our harbors are a forest of merchant ships, and the people die of hunger."[14] These "extremities of human degradation" were not accidental; on the contrary, said Ruskin with such sharpness that his books could no longer find a publisher, they were due "to the habitual preying upon the labour of the poor by the luxury of the rich." But it was not only the physical and material condition of the people which alarmed him. What was happening to their artistic and creative genius? The rationalization and specialization of industrial processes were destroying it. In the factories the craftsmen of former times were "divided into mere segments of men — broken into fragments and crumbs of life." At home they were surrounded by a flood of machine-made articles in which Ruskin could see no beauty — the "paltry art" of the Crystal Palace — while the rich monopolized "the music, the painting, the architecture, the hand-service and horse-service, the sparkling champagne of the world."

Fig. 104. The Crystal Palace, London, 1851. Joseph Paxton, architect. Ruskin's hostility to industrialism blinded him to the significance of this first great monument to its architectural potentials.

What was to be done? Ruskin's distrust of the machine was equaled only by his distrust of Victorian democracy. He was not only a Tory in British politics; he was an aristocrat in principle. For all his own *nouveau riche* origins, he was unashamedly the spokesman for the nobility, the landed gentry, the gentlefolk of England whose manners, morals, and esthetic standards were now being eclipsed by Victorian capitalism. Where the arts were concerned, he saw it as only proper that "the decision [be] made, the fame bestowed, the artist encouraged . . . by the chosen few, by our nobility and men of taste and talent." As a parallel, it is not surprising to find that Ruskin, at first merely ignorant of science, became in his later years profoundly anti-scientific.

Against these characteristics — anti-democratic, anti-industrial, anti-scientific — was opposed Ruskin's genuine humanity, his deep attachment to the individual man, whose talents were being aborted, honesty deflowered, and happiness destroyed. On this paradox turned all his arguments and rested his final decision. Since all theories of progress were fallacious, he could not go forward, he could only turn back. And only medieval society, the "magnificently human" Gothic styles of Europe, could afford him that balance which he sought between security and discipline.

Stripped of all rhetoric, this was a program of political, social, and cultural reaction. To begin with, his picture of Gothic life was historically inaccurate, as even casual reference to original medieval sources should have told him. And to propose that the productive forces of the British Isles be reorganized at the level of King John and the Magna Carta was sheer escapist nonsense. This was equally true, whether Ruskin discussed political economy or architecture. He himself made them indivisible: "*The Stones of Venice* had, from the beginning to end, no other aim than to show that the Gothic architecture of Venice had arisen out of, and indicated in all its features, a state of pure national faith and of domestic virtue."[15] In building design, as in political economy, his program was reactionary: for objective reality made it certain that the schism between technique and ideology would not be resolved by a mere buttering of the Gothic or the Classic or of any other historical style.

HORATIO GREENOUGH, YANKEE DEMOCRAT

Of all the men who worked within the field of formal criticism, none was so advanced as the Yankee stonecutter, Horatio Greenough.

He died young, at the age of forty-seven, leaving only a small and fragmentary collection of essays behind him. He was well known to only a small group of his contemporaries and his influence seems to have been totally submerged by his premature death. But had his essays circulated as widely as those of Ruskin, American design might well have escaped the decades of tortuous frustration which lay between Greenough and Louis Sullivan. For Greenough, at the very dawn of industrial production, anticipated to an extraordinary degree the problems it would raise in design.

In depth and comprehensiveness, Greenough's analysis of technical, political, and esthetic factors in design have scarcely been equaled by an American.[16] This grasp of critical values is all the more surprising in a man who was a sculptor all his life — and not a

Fig. 105. "George Washington," 1832–41. Horatio Greenough, sculptor. This heroic sculpture, designed for the rotunda of the U. S. Capitol, created a furore because of its nudity. Exiled first to the Capitol grounds, it now rests in the Smithsonian.

Fig. 106. "The Rescue," 1837–51. Horatio Greenough, sculptor. This monumental group was designed for the West Front of the Capitol where it stood until recently.

great one, at that. (He himself had feared that he would never realize "my idea of what sculpture should be.") But he clearly had sharpened his wits on Italian marble. He had begun as a student of the Classic, adoring Nature, "but as a Persian does the sun, with my face to the earth." He ended by studying Nature, with his eyes full on her face.

Horatio Greenough was an exponent of Boston at her greatest. Graduated from Harvard in 1825, he had trained in Boston's Athenaeum in the heydey of the Greek Revival. He had taken so literally the enthusiasm for the rediscovered Classic world that he had gone abroad to be at its fountainhead. In Italy he had been a member of the large and active colony of American intellectuals during the exciting times of Mazzini. While there he had executed the famous statue of Washington — the seated General clothed only in a toga, a blasphemy which an outraged Congress would not permit in the Capitol (Fig. 105). It was in the controversy over this commission that Greenough had to take to the pen in order to explain and defend his concepts of art.

He had spent much of his adult life in Italy. But in 1851, with the Italian democrats scattered in exile, he had returned for good to his native land. It was immediately apparent that Greenough had been no lavender-tinted expatriate. He had crossed the Atlantic six times, not to escape the American, but to be closer to the Classic, mind. And when he came back for good, it was clear that he *had* penetrated to the very heart of the matter. If American architects were to profit from Greek experience, they must study it "like men and not ape [it] like monkeys." If they wanted to achieve truly satisfactory esthetic standards, they would have "to learn principles — not copy shapes." And what were these principles, where were they to be observed? Why, "the manly use of plain good sense" would reveal them in every facet of American life: the trotting-wagon, the New England farmhouse, the clipper ships, the lighthouses and canals (Figs. 109, 110).

He strode across Manhattan — become in his absence America's greatest metropolis — and his sharp, buoyant eye was quick to see the contrary tendencies in the construction all about him. Isaiah Rogers's new Astor House, with a Classic façade for Broadway and

Fig. 107. Trinity Church, New York, N. Y., 1839–46. Richard Upjohn, architect.

Fig. 108. The Reservoir, New York, N. Y., 1855. James Renwick, architect.

only naked masonry for the rear. Richard Upjohn's scholarly Trinity Church in Jersey brownstone. There were not one but two Americas here, said Greenough, and "the puny cathedral of Broadway, like an elephant dwindled to the size of a dog, measures her yearning for Gothic sublimity while the rear of the Astor House and the mammoth vase of the great Reservoir show how she works when she feels at home."[17]

Significantly, Greenough did not ally himself with any of the quarreling schools of stylistic controversy. He was convinced that no eclecticism would suffice. Versailles was a "royal mushroom" and Barry's new House of Parliament just then building in London was nothing but a "gewgaw." On the other hand, he disapproved of Jefferson's use of a Roman temple as the prototype for the Virginia capitol; and the dark Romanesque mass of James Renwick's new Smithsonian Institution "scared" him. What was the "castle of authority . . . outwork of prescription . . . fortress of theology" doing in the capital whose proudest ornament was the Bill of Rights? Was there "no *coup d'état* lurking" in those Lombardian masses?[18] He asks himself: "Is it possible that out of this [eclectic] chaos order can arise? — that of these conflicting dialects and jargons a language can be born? When shall we be done with experiment? What refuge is there from the absurdities that have successively usurped the name and function of architecture?"[19]

Like Walt Whitman, but with much more precision, he saw in the burgeoning American industry the material basis for a new esthetic standard, one which would be in conformity with advancing science and technology instead of increasingly in opposition to it. Here his confidence in the creative genius of the plain people was not vitiated, as was Ruskin's, by a Tory's fear of giving them too much freedom. It was they who would fashion the new American beauty — not the intelligentsia or the artists or the wealthy and well-bred connoisseurs. "The mechanics of the United States have already outstripped the artists and have, by their bold and unflinching adaptation, entered the true track and hold up the light for all who operate for American wants."[20]

What kind of beauty could we expect of them? Greenough an-

Fig. 109. The yacht "America," which won the Queen's Cup at Cowes, England, on 22 August 1851, exemplified the kind of anonymous design by skilled craftsmen that Greenough praised.

Fig. 110. The fast trotting wagon or sulky was another American artifact which was famous for its functional elegance and lightness.

swered with poetic and electrifying precision: "By beauty I mean the promise of function. By action I mean the presence of function. By character I mean the record of function." And where could examples of such beauty be found? Observe the Yankee clipper ships, which had already caught the imagination of the country:

> Here is the result of the study of man upon the great deep, where Nature speaks of the laws of building, not in feather and in flower, but in wind and waves, and he bent all his mind to hear and to obey. . . . If this anatomic connection and proportion has been attained in machines and — in spite of false principles — in such buildings as make a departure from it fatal, [such] as bridges and scaffolding, why should we fear its immediate use in [building] construction?[21]

It is apparent that Greenough had given deep and independent thought to the crisis in building design. Just as he refused to copy the Greeks, so he refused to imitate the shipbuilders. It was principle he was after, not prototypes.

> Here is my theory of structure: a scientific arrangement of spaces and forms [adapted] to functions and to site; an emphasis of features proportioned to their *gradated* importance in function; colour and ornament to be decided and arranged and varied by strictly organic laws, having a distinct reason for each decision; the entire and immediate banishment of all make-believe.[22]

Greenough was no crass materialist, naïve enough to suppose that any product of man could be empty of esthetic significance; but he was searching for a new set of standards which would not be imposed from the top, and from the top of the past, at that. These standards would appear naturally, as an orderly exponent of the very process of creation. Yet, while they would be a logical — in fact, an automatic — by-product, they would not be accidental or intuitive. He stipulated that they develop within a scientific reference frame. He thereby avoided the pitfall of the Romanticists. He was under no illusion that Rousseau's "natural" man, merely left to his own devices, will finally produce a satisfactory building. He was discussing flesh-and-blood American workmen, rapidly learning that folk-knowledge was not enough for industrial production; that the laboratory was the womb of the factory, science the seed of design.

There was, as Emerson pointed out, a similarity between Ruskin and Greenough, "notwithstanding the antagonism in their views of the history of art." This similarity lay in the fact that they both recognized the impossibility of discussing esthetic standards as isolates. Such standards were specialized expressions of society, complex and intricately related to every aspect of social life; but susceptible to analysis nevertheless. And if one wished to discuss or compare or alter them, one had always to relate them to their respective social orders. On this the two men were in agreement; but never did more divergent conclusions flow from identical premises. It would be futile to analyze this antagonism without reference to the class and national platforms from which the two men spoke.

Ruskin had begun as the ideologist of the old gentility. He had clashed sharply with the voracious appetites of the capitalists. He was appalled at the hypocrisy and ugliness of Victorian England. Yet Ruskin's opposition was always limited. He gave reaction some of its most persuasive arguments for turning the clock back. His mechanistic analogies between pious intentions and artistic success were quite as fallacious as Pugin's, and far more insidious because far more persuasively put. His hysterical interjection of moral judgements into controversies which were essentially esthetic contributed measurably to the triteness and confusion of the period. Despite his real concern for the state of the working man, he had no confidence in that man's capacity to make any decisions for himself. Though he did indeed pour a good portion of his personal fortune into schemes of his own devising for social uplift, these schemes were basically feudal in inspiration. Unlike William Morris, he was incorrigibly hostile to trade unions, Socialism, democracy itself. It is one of the least attractive expressions of his imbalance that he actually found it possible to defend slavery in the American Confederacy.

His separation of esthetic from technique was, objectively, a reactionary maneuver. *Stones of Venice* or Houses of Parliament, book or castle, these creations were products of literature, not life; propaganda, not building. And the fact that *The Stones* would not protect an Englishman from a rainstorm quite as effectively as the House of Commons did not alter their basic similarity; they were both instruments of persuasion, designed to sell the British people the beauties of medieval serfdom. Ruskin was merely one cog in the giant obscurantist machine which was overwhelming whole sections of English culture.

Greenough, too, was the ideologue for an upper class, but with this difference: he came from the heart of Yankee industrialism at the very moment it was preparing to challenge the slave oligarchy for the mastery of the nation. Greenough had grown to manhood in the great Boston of the Renaissance. He had studied Greek art at the Athenaeum and thrilled to the Greek Wars of Liberation. Dr. Parkman had given him lessons in anatomy, Washington Allston and Richard Henry Dana encouragement. In Paris he had done a bust of the aging Lafayette. In Italy he worked for nearly eight years on the ill-fated figure of Washington; there, too, he had seen the struggle for Italian independence under Mazzini. When Greenough spoke, therefore, he voiced the aspirations of the most progressive section of mid-century American democracy. It is this reality — democracy, science, industry — which he accepts as the essential basis for effective esthetic standards. This reality is the source of his buoyant confidence, enabling him to envisage a democracy so complete as to render esthetic falsehood impossible. His buildings would have no ulterior esthetic motive — no intangible job of intimidation, repression, or deceit — precisely because the society of which they were a part would have no such motives.

Fig. 111. Sculpture Gallery, The Athenaeum, Boston, Mass., in 1855. At the time of his death, the scene of Greenough's earliest studies boasted one of the nation's largest collection of sculpture and casts.

5.

THE GOLDEN LEAP

The nineteenth century saw three great developments in structural theory: the enclosure of great areas in the Crystal Palace; the spanning of great voids in the Brooklyn Bridge; and the reaching of great heights in the Eiffel Tower. These structures are sufficiently great in themselves, each marking the full-flowering of a new structural concept. But they are noteworthy in another respect in that they constitute historical proof of the relationship between structural theories, materials, and techniques. The interaction between these three is constant and dynamic. Together, they constitute what I have called the technological level of building. Their relation is such that an advance in one naturally and inevitably affects the other two; and this reciprocal action is the mainspring of evolutionary development.

Historically, the three factors — theory, material, technique — have seldom been in exact equilibrium at any given moment: that is, there have been few times when a lag in one did not prevent advance in the others. Occasionally, however, under the accumulating pressure of social change, a structure appears in which all three factors have combined at a high level to produce a radically new type. To borrow a term from the anthropologists, there has been a *leap* forward. Such structures were the Palace, the Bridge, and the Tower. These leaps involve not merely blind social and economic pressures. Specific human agencies are also required; live men who, by the very breadth of their understanding, are able to master all the factors involved and force the project through to completion. Such men were the designers of these three famous structures: Joseph

Paxton, the English horticulturist; John Roebling, the German-American cable inventor; and Gustave Eiffel, the French engineer. The shadows of these men and their structures fell athwart the whole Western world; and nothing which came after them could remain entirely unaffected, for they had tackled and brilliantly mastered three eternal problems in structure.

Paxton, Roebling, and Eiffel were men of the new world of nineteenth-century science and thus closer to us who follow than to those who preceded them. Together they constitute an organic continuum which spans both the century and the century's chief centers of intellectual activity: London, New York, Paris. Their thinking and their works were truly supranational in effect. Like Darwin and Marx, Morgan and Pasteur, they were citizens of world thought, at once the creators and the first great products of modern scientific theory and practice. They differed from their predecessors in this important respect: they not only worked in the field of science; they had also begun to think scientifically. Jefferson's fertile intellect might be stimulated by contemporary investigations in natural phenomena; Fulton, from God only knows what accidental contacts, might putter around with his steam boiler until the boat finally propelled itself; and Franklin might risk the thunderstorm with his kite and key. Yet they lacked in their investigations the methodology, the rigid standards, the planful accumulation of fact, which characterizes modern science. Theirs was not yet the world which, as Paxton himself put it, "requires [of the scientist] facts instead of empiricism, and prefers scientific accuracy to blind practice."

THE PALACE

Of the three men, Paxton (1803–1865) was the least well educated and, as a consequence, the closest to the methodology of the preceding century. Self-made in the strictest sense of the word, Joseph Paxton began the first of a series of apprenticeships on the great country estates of England at the age of fifteen. Under ordinary circumstances, this training would have led to the status of head gardener, but both his aspirations and abilities were much higher. The London Horticultural Society had recently leased the gardens of the Duke of Devonshire at Chiswick and begun there a program of reconstruction and improvement; and what was to prove the turning point in Paxton's whole life came when, in 1823, he obtained employment there.

Work for a learned society offered him much greater perspectives than gardening for a country gentleman. The whole direction and emphasis of the Society lay in extending and organizing botanical and horticultural knowledge, rather than in mere competitive display. Young Paxton throve astonishingly in this environment. In three years he was a foreman at Chiswick, though his salary was a mere eighteen shillings a week. He was apparently confident that he could do much better and was on the point of sailing for America when the same Duke of Devonshire, now president of the Society, offered him the post of superintendent at his country seat, Chatsworth.

Because of the wealth and prestige of the Duke, this offer could scarcely have been topped by any but the royal house itself. Devonshire was one of the leading patrons of horticulture; in addition he seems to have been an ideal employer, for he allowed Paxton great freedom and ample funds. (Paxton confirmed this years later, in 1851, at a dinner in his honor after the Crystal Palace was opened: "By his confidence and liberality I have had placed before me ample means for various experiments without which there would never have been a Crystal Palace.") During his long association with Devonshire, Paxton achieved the status of both gentleman and scholar. In the former capacity he made the Grand Tour of Europe and Asia Minor with the Duke; in the latter, he became author of several books and innumerable articles as well as editor of the *Magazine of Botany and Register of Flowering Plants.* This periodical appeared first in 1834, handsome in makeup and professional in content. The very first issue reflected the curious conflict in the lowly born Paxton between snob and scholarly democrat: he dedicated it with flowery servility to the Duke. (Each succeeding volume was similarly dedicated, and never to anyone less than a duchess. Since the magazine must of necessity have been subsidized, however, we can understand these archaic attentions to the patrons.) But in his introduction he pledged himself to keep his material "as plain and intelligible as possible," avoiding Latin and botanical terms except where absolutely necessary, so as to make the magazine useful to the widest possible audience.

The pages of his *Magazine* alone are evidence of the reach and depth of Paxton's intellect. Since much of the horticultural work on the great English estates revolved around the production of exotic and/or out-of-season fruits and flowers (pineapples, bananas, and oranges; winter melons and grapes; rare orchids and tropical lilies), it is not surprising to find so much of Paxton's attention focused on techniques for manipulation of the unfriendly English climate. Thus, in a paper titled "Influence of Solar Light on Vegetation," he anticipates the discovery of the ultraviolet energy band, speculating that there must be "rays issuing from the sun [which are] distinct from the rays of heat and light — rays of chemical and magnetic influence; and who can tell to what extent glass may not intercept and transmute these rays?"[1] He experimented with the use of colored glasses, testing their value as filters for this mysterious property. He tilted his glass roofs at different angles, to bring them nearer to 90 degrees with the low, weak English sun. And for the same ends of more precise environmental control, he made himself the master of heating and ventilating techniques. He reported on extensive experiments in steam and hot water heating, in improved combustion and smoke abatement. He even designed a mechanical stoker for his boilers.

But his practical work, during the 1830's and 1840's, fully matched the theoretical. Although he was increasingly active as a designer of gardens, parks, and new suburban developments outside of Chatsworth, Paxton carried on a steadily expanding program of glasshouse construction.[2] Thus, between 1828 and 1850, he built at least eighteen greenhouses for palms, melons, orchids, and water plants (twelve at Chatsworth, six elsewhere). These show a consistent development toward more transparent, more waterproof and stronger glass construction. He gradually perfected his famous wooden glazing system called "ridge-and-furrow." Like many pragmatic inventions, this system simultaneously solved several disparate problems with an uncanny simplicity of means. It had structural members at each furrow which also served as gutters to carry both

Fig. 112. The Great Conservatory, Chatsworth, 1836–39. Joseph Paxton and Decimus Burton, architects. Here Paxton used his ridge-and-furrow technique on an elliptical roof 277 ft. long, 123 ft. wide, and 67 ft. high, to house exotic trees and plants of largest size (left, below). Demolition begun (right).

Fig. 113. *Victoria Regia* Lily House, Chatsworth, 1849–50. Joseph Paxton, architect. The second house designed specifically for the lily: here it bloomed, thanks to carefully warmed air, water and even mud at roots.

Fig. 114. Capesthorne Hall Conservatory, Cheshire, c. 1837. A gently-pitched roof of ridge-and-furrow glazing is carried by semicircular arch ribs of laminated wood. Columns and trusses were cast iron.

rainwater and interior condensation into the hollow cast iron columns which were the vertical support. It used the new large sheets of rolled glass (9″ × 36″) in a chevron pattern which further strengthened the corrugated section of the ridge-and-furrow: these combined to make the system highly waterproof and wind-resistant. Having perfected this system, he proceeded to apply it to buildings having lean-to, gabled, and flat roofs. Then, in a dramatic demonstration of its further potential, he used it in the vaulted dome of the Great Conservatory at Chatsworth in 1836. Here, by *bending* the ridge-and-furrow at right angles to its corrugation, he achieved an astonishing glass shell, as stable as it was beautiful (Fig. 112).

But, whatever else he was, Paxton had become, by 1849, the leading gardener of the realm. For in that year he was successful in bringing to flower, for the first time in Europe, the famous water lily *Victoria regia.* The lily was an exotic, a native of equatorial South America, but though it had been grown in many tanks, it had never bloomed. To bring it to flower was a problem of the most precise environmental control, as Paxton fully understood. He would have to house it in a new building especially designed to duplicate its natural habitat. And this was no easy matter: "such plants require more light than our glass will transmit and yet more heat than our open sky affords, and are consequently most difficult of culture . . . They require a [glazing system] that transmits all possible light."[3] He built a new house, employing a flat roof of ridge-and-furrow, and an elaborate system for heating the air, the water, and the earth at the bottom of the tank. He provided for draftless ventilation and a gentle flow of water through the tank. By a careful manipulation of all these factors, *Victoria* grew and bloomed. "Scientific accuracy and not blind practice" had enabled him to duplicate the natural circumstances of the plant, "considering [its] wants, whether as regards light, heat, air, moisture, or soil."

Two years later, when the Palace opened its doors to a dazzled world, it was apparent that Paxton had studied another aspect of *Victoria regia:* its structure. This remarkable plant had pads which measured eight, ten, even twelve feet in diameter: constructions strong enough to support Paxton's young daughter yet light enough

Fig. 115. Transept, Crystal Palace, Hyde Park, London, May 1851. Joseph Paxton, architect. The splendid crystal vault was devised to clear a group of fine old elm trees whose preservation was mandatory. The flat-roofed nave (below) carried the typical ridge-and-furrow roof to a total length of 1848 ft.

to float. They were marvels of structural economy — a network of radial ribs tied together by circumferential purlins which supported, but were in turn stiffened by, the thin tough membrane of the leaf itself; and the whole unit further strengthened by the upturned, scalloped rim. The parallel between this design and that of the Palace itself is not fortuitous. Indeed, as Paxton said himself, "Nature was the engineer. Nature has provided the leaf with longitudinal and transverse girders and supports that I, borrowing from it, have adopted in this building."[4]

The Palace in Hyde Park created an immediate sensation. The light and airy framing, the glittering curved vaults of glass soaring over full-grown forest trees, the fountains and tropical flowers, the printing presses, pianos, sewing machines, and sculpture — this juxtaposition enchanted the Victorian temperament. By some miracle of history, the first great exhibition glorifying modern industrialism was housed in a structure which more perfectly expressed its potentials than any that ever followed it (Fig. 104, 115).

The building was one of great beauty and many extraordinary aspects, not least of which were the circumstances surrounding its design and erection. Paxton, as is well known, won the design after the competition had closed and some 233 other designs, including the winner, had been rejected. His office at Chatsworth completed the working drawings in *eight days* of continuous labor. It was the first completely *prefabricated* building (as it was the first *demountable* building) of modern times. Composed of standardized components — all of them based on a basic 24-foot module — it was easily bolted together and subsequently dismantled under Paxton's guidance, who re-erected it in Sydenham in 1852–54. It was far and away the largest single enclosed volume the world had seen: 989,-884 square feet of floor space, one and one half miles of balcony, 17¾ acres of Paxton's ridge-and-furrow roof. About 900,000 square feet of glass went into it, along with more mass-produced iron columns, beams, and girders than ever seen before. Its erection was a miracle of speed: the contract signed only on October 31, 1850, it was opened to the public on May 1, 1851. The speed was largely due to the fact that site work consisted almost exclusively of the *assembly* of the fabricated parts; all actual *fabrication* took place under controlled conditions in factories elsewhere.

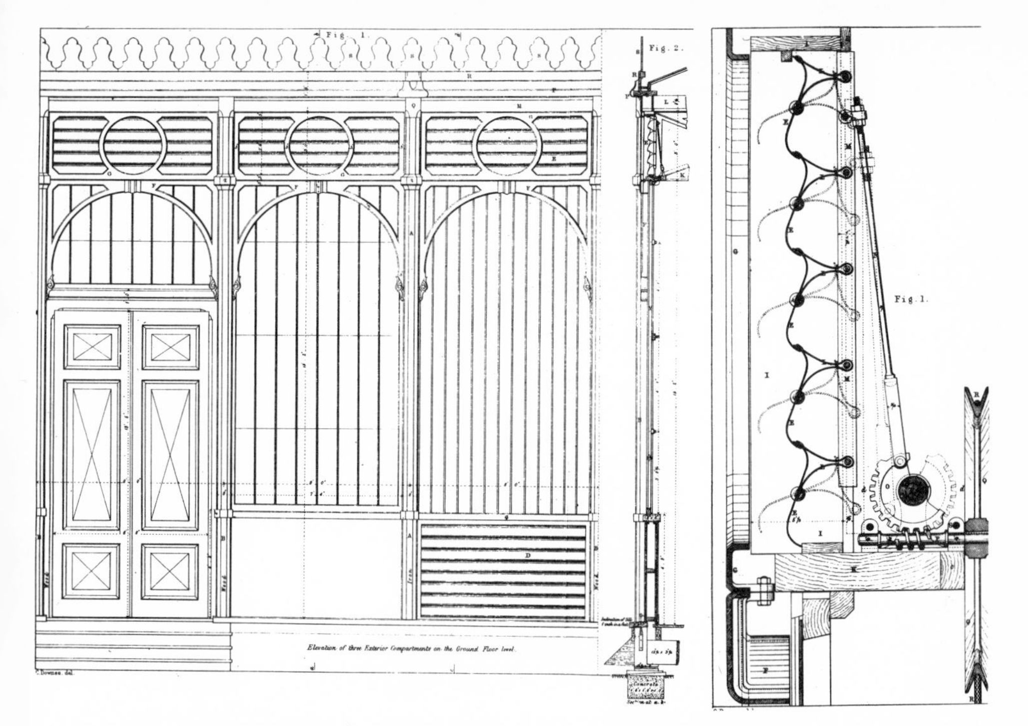

Fig. 116. Exterior wall, elevation and section. Three variations of the standard 8 ft. wall panel were provided: double wood doors with glass transom; fixed glass with solid wainscot; solid wood boarding with louvered wainscot. All had transoms of operable metal louvers (detail, right).

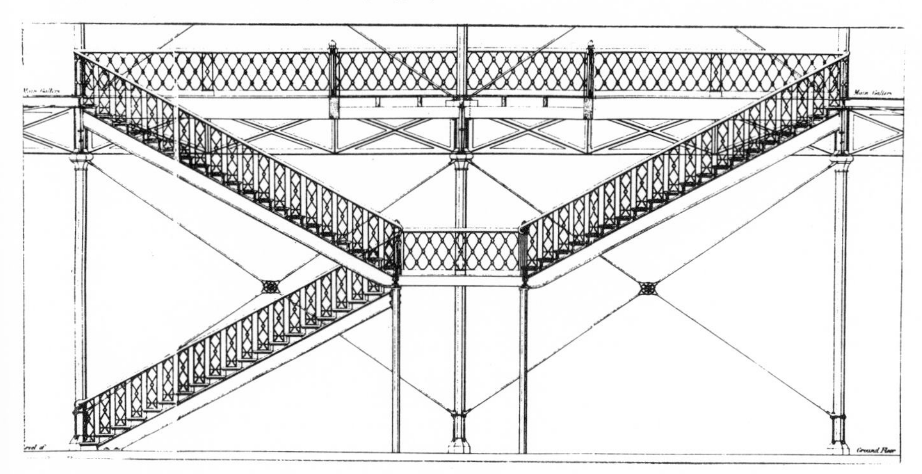

Fig. 117. Lateral stability was given frame by standardized diagonal bracing. See details below.

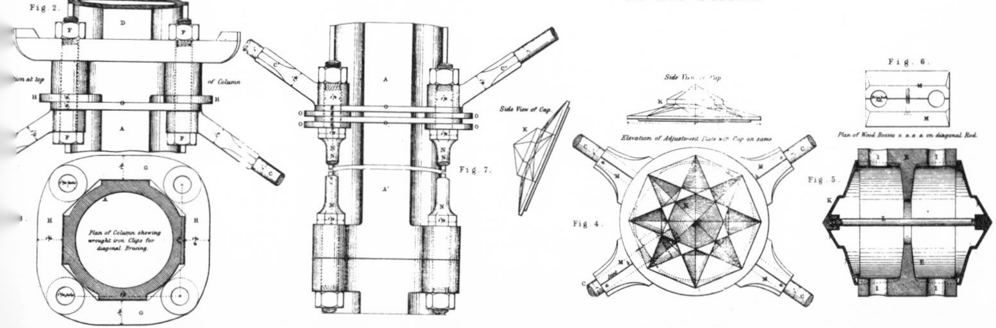

But none of these statistics succeeds in conveying the full dimensions of Paxton's accomplishment at Hyde Park, for the Crystal Palace really introduced a whole new order of magnitude into building performance. With it appears, for the first time, the modern concept of *strength through precision instead of sheer mass.* By his separation of building tissue into skeleton and skin, he was able to enclose enormous volumes with an integument of relatively little weight or thickness. He may have learned this lesson from organic forms but he did not make the typical Victorian error of trying literally to mimic nature: rather, he mastered her principles. Unlike his lily, his structure had no water upon which to float. To give it stiffness, he therefore corrugated it into his ridge-and-furrow, creating a membrane of great stability even when used in a single plane. When he bent it into a semicircular vault he was, in fact, doubling the corrugation effect. And this gave him the soaring transept at which, quite properly, the Victorians gasped.

Joseph Paxton's investigations of structural phenomena, flowering in the Crystal Palace, must therefore rank as one of the great scientific victories of the century. For a man with little or no formal education, they are all the more remarkable. It is true, of course, as Chadwick has recently demonstrated,[5] that Paxton had able assistants in the design and erection of the Palace: the engineer William Barlow, who calculated the required strength of columns and girders; the ironmasters Messrs. Fox and Henderson, who acted as both fabricators and general contractors; Owen Jones, who evolved the lovely color scheme of pale blue skeleton with yellow chamfers and bright orange-red flanges. But it is also obvious that, in all important respects, the design is Paxton's — the result of over two decades of intensive work in glass construction.

The Crystal Palace marked the apex of Paxton's career. Already a wealthy Member of Parliament and respected member of his profession, he was now knighted by his delighted Queen (who summoned him to a private audience and paid him the unprecedented compliment of standing behind her sofa throughout the entire interview). Though some of his better-trained colleagues may have been jealous of his success, they were drowned out by popular ac-

Fig. 118. General view, The Crystal Palace, Hyde Park, London, 1851. Joseph Paxton, architect; Fox and Henderson, engineers and general contractors.

Fig. 119. The light, modular frame was completely fabricated in the shop and assembled on the site. Result was unprecedented speed in erection, tidiness of construction site, safety for workers.

claim. Even the hostile Ruskin was forced to grudging praise after the Palace was re-erected at Sydenham, though he wondered "why the poetical public insist on calling it the Crystal Palace [when] it is neither a palace nor of crystal." He admitted that it was better than the "paltry art" it contained, expressing "a single and very admirable thought of Sir Joseph Paxton's . . . that it might be possible to build a greenhouse larger than ever greenhouse was built before. This thought, and some very ordinary algebra, are as much as all that glass can represent of human intellect."[6]

The Palace gave Sir Joseph an international fame, entrée into the highest circles and as many commissions as he and his architect son-in-law could handle. But he was never to excel this high point of his career. Even the Sydenham reconstruction became, in the very process of elaboration and enlargement, coarse and more mannered than the original. Paxton designed several more crystal buildings. One of them, the proposed pavilion for the Paris suburb of St. Cloud, would have been even more breathtaking than the original. But none of these materialized. His conventional architecture, on the other hand, continued as before to reflect the taste of the period. It is almost as if there were two Paxtons at work here, with an impermeable barrier between them: the discontinuities between his glass houses and his great residences, like those for the Rothschild family at Mentmore and Ferrières, are startling.

The Crystal Palace, then, must be regarded as the flowering of Paxton's special genius in a peculiarly fortunate set of circumstances — the same sort of event, in fact, as the flowering of *Victoria regia*. These circumstances were: a concrete and specific program calling for continuous, well-lighted floor space; sharp limits to the funds and time available; the suitability of the structural system on which he had already spent so much time and energy. Even the eight days allowed for preparing the drawings were auspicious. They allowed no time to worry about appearance, so that the finished building shows a beautiful innocence of current architectural controversy over idiom and cliché. As a result, the building was as lean and functional as a greyhound, revealing only in its smallest details (column caps, tie bars, brackets) the imprint of Victorian taste. In his later work these controlling factors were missing and without them

Fig. 120. Mentmore, Buckinghamshire, 1855. Joseph Paxton and G. H. Stokes, architects. This great house, one of several for the Rothschilds, shows Paxton's completely conventional response to standard "architectural" problems.

he was lost — a self-made man in an overawing environment.

One of the enthusiastic assumptions of the Victorians concerning iron construction in general and the Palace in particular was destined to prove quite fallacious — that is, that a fireproof structural system had at last been perfected. Fire had always been one of society's greatest hazards and fire resistance had naturally been a most sought-after property in buildings. Unfortunately, it was seldom achieved. Until the appearance of mass-produced metals, wood had been the only building material with high tensile strength. All but a microscope proportion of nineteenth-century buildings employed floor beams and roof trusses, and these were necessarily of wood. Thus, when mass-produced metal columns and beams appeared in mid-century, they were uncritically acclaimed. This enthusiasm was based upon the fact that neither iron nor glass will burn. It conveniently overlooked the corollary that both will melt. Under a given set of conditions, a wooden beam is more firesafe than a steel one: the steel will deform and collapse under a heat which only chars the wood. To reduce the strength of a wooden beam, you have to reduce its cross section. Paxton's structure served as inspiration for an inexpert copy in New York City in 1853 (Fig. 122). It was completely destroyed by a twenty-minute fire, and in 1936 the same fate overtook the Crystal Palace itself.

Fig. 121. Crystal Palace (proposed), New York, N. Y., 1853. James Bogardus, architect. Central tower 300 ft. high supports sheet iron tent carried on iron catenaries attached to exterior walls of circular amphitheater. Entire 700 ft. structure was to be of modular, prefabricated cast iron elements.

Fig. 122. The Crystal Palace, New York, 1853. Carstensen & Gildemeister, architects. A 20 minute fire destroyed it in 1858, thus proving that iron unprotected would melt quicker than wood would burn.

Fig. 123. Main Exhibition Building, Centennial Exposition, Philadelphia, Pa., 1876.

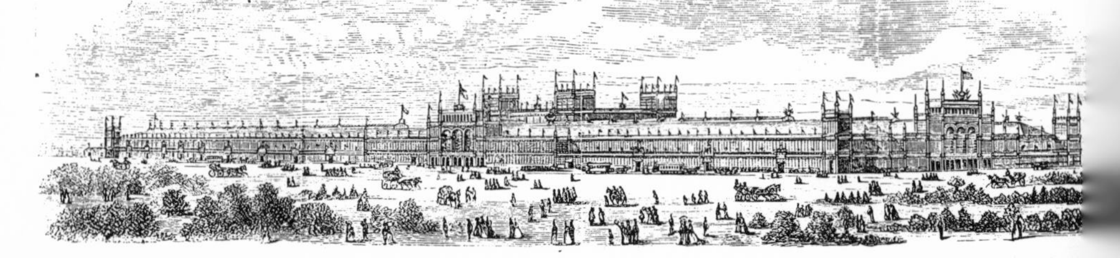

THE BRIDGE

Exactly twenty years before the Crystal Palace, John Augustus Roebling (1806–1869) had migrated to Pennsylvania. Neither the time nor the place was accidental. The coal fields were roaring into action, Pittsburgh was already the center of a new iron industry, and canals hardly completed were proving themselves inadequate to handle the freight. Very shortly a network of railroads would be flung across the Alleghenies. The bridging of thousands of mountain streams would become a pressing problem. There would be neither time nor money for elaborate masonry structures — even had there been a masonry tradition in this country, which there was not. In addition, the size of the rivers, with their ice floes and turbulent spring freshets, made unobstructed spans a necessity. Nor could the huge cantilevered trusses appear until after mass production of standardized steel members began in the rolling mills of the seventies and eighties. In such a context, Roebling's experiment with tension structures was inevitable.

John Roebling did not, of course, invent the suspension bridge. On the contrary, as an engineering student in Germany, young Roebling had made the chain suspension bridges of Germany and Switzerland the subject of his thesis. Neither did he actually invent steel rope. What he did was to force the parallel development of both bridge and cable. Almost alone, for over thirty years, he forced their simultaneous improvement in a series of increasingly spectacular suspension structures. Theory, material, techniques: each felt the impact of his intellect. The potentials he saw in the suspension bridge could only be exploited by a tensile material far superior to hempen rope or hand-forged chain. He achieved the first wire cables; but these in turn implied new fabrication and assembly techniques. Having perfected the latter, he could then turn his attention to still more daring designs. This ascending spiral reached its most polished statement in the Brooklyn Bridge — a design which has never been surpassed (and only equaled in such bridges as the Whitestone and Verazzano in New York and the Golden Gate in San Francisco) in its ultimate simplicity.

There is certainly nothing surprising in Roebling's accomplishment. His entire career is one of amazingly consistent preparation for his final design.[7] The initial decision to migrate to this country was a reflection of his dissatisfaction with the prospects which Europe offered a young engineer. He and his brother had made a careful study of the States before deciding upon Pennsylvania as a future home. They rejected the South because of their bitter opposition to slavery. It was, John felt, "the greatest cancerous affliction . . . enough for us not to go into any slave-holding state, even if Nature had created a Paradise there." Although they had initially bought a 7000-acre farm, John Roebling soon (1837) left it for engineering.

His first employment was on the Pennsylvania Canal, on that remarkable section where, by a system of locks and inclined railways, the barges were carried over the crest of the range. In this admirable system, the weakest links were the hawsers which pulled the barge-laden cars up the inclines; although they were woven of best Kentucky hemp, they were continually breaking. Why could these cables not be woven of wire instead of hemp? To answer the question was, with Roebling, a necessity. He set to work designing the machinery for weaving such a rope and in 1841 perfected it. This first cable was literally a steel rope composed of small spirally twisted wires. Roebling was quick to see the potentials of his new material: the availability of steel cable had the immediate effect of liberating the suspension structure from the limitations of hand-forged chains. In 1845 he completed his first suspension structure, an aqueduct for the Pennsylvania Canal. A structure quite without precedent in the New World, the aqueduct consisted of seven spans of wood flume, each one hundred and sixty-two feet long, carried by two continuous cables seven inches in diameter.

But even before he had the opportunity to erect a suspension structure with his twisted rope, Roebling had perfected a second and even more significant cable. Unlike the first, the individual strands were *parallel* (instead of twisted into a spiral). This meant that each individual strand would be identically stressed. Small as it seemed, this change immensely improved the efficiency and pre-

Fig. 124. Inclined Plane, Mt. Pisgah, Pa. (left) Inclined Plane on Morris Canal (right). Expensive hempen cables used on such projects convinced Roebling of need for perfecting steel wire rope.

Fig. 125. Niagara Railway Suspension Bridge, Niagara, N. Y., 1851–55. John A. Roebling, engineer.

Fig. 126. The Ohio River Suspension Bridge, Cincinnati, Ohio, 1856–66. John A. Roebling, engineer.

dictability of suspension structures. Roebling himself used it in all his larger bridges and it has been a standard procedure ever since.

Roebling could easily have become a manufacturer exclusively, but his interest in design was now thoroughly aroused. Building a new factory in Trenton (1848) did not prevent his building a new highway bridge across the Monongahela at Pittsburgh in 1846 (eight spans of one hundred and eighty-eight feet each) or the railroad bridge across Niagara Falls in 1854. Indeed, as the size of the structures increased, new problems arose. To reduce the danger of failure it was necessary to take over the manufacture of the wire from which the cables were woven; this involved improved metallurgy. And for such great structures the cable could no longer be woven in the Trenton plant. In the Pittsburgh Bridge, the cables were woven on the bank. Ultimately, on the Brooklyn Bridge, even this technique would not work. Again the man was equal to the job. He evolved a traveling weaving machine, which shuttled back and forth across the East River, on temporary cables, weaving the main supports as the spider does its web.

All this did not exhaust Roebling's wide interests. In 1847 he advocated a system of railroads and telegraph lines to replace the canals and horse couriers. In 1850, four years before Cyrus Field became actively interested, he wrote that a trans-Atlantic cable was entirely feasible, even giving his specifications and estimate of its cost. Then, in 1857, with two bridges under construction at Cincinnati and Pittsburgh, he wrote to Abram S. Hewitt of New York City proposing a bridge from Manhattan to Brooklyn. The proposal caught the imagination of New Yorkers, but preliminaries were slow and the Civil War intervened. His son Washington went off to war with the Union army and the plant at Trenton was converted to war work. Thus it was not until 1869 that the project went through. Roebling's design was accepted and Roebling appointed chief engineer.

As much a part of American life and landscape as Niagara Falls, the structure needs no description here (Fig. 127). The scale and breathtaking clarity of its design are indelibly imprinted upon American memories. But more than this must be said if the full stature of Roebling and his bridge are to be understood. The pas-

Fig. 127. The Brooklyn Bridge, New York, 1867–83. John A. Roebling, engineer.

Fig. 128. Construction Views.

sionate attachment to the work, which cost him his life in the first year of construction and cost his son his health in the next, was but one aspect of Roebling's mastery of the field. For, in addition to manufacturing the cable, perfecting the machine which wove it, training the workmen and supervising the overall construction, he conceived the design itself. He thus proved himself the master of the large as well as the small. Posthumously, he displays a complete understanding of tension structures. He established criteria which are still operative; and much of what has been added subsequently reveals itself as either ostentation or ignorance.[8]

THE TOWER

Perhaps the most dramatic leap of the century was that of the Eiffel Tower. Unlike Paxton and Roebling, Gustave Eiffel (1832–1923) met the most determined — one might almost say political — opposition. He had to fight not only technological lags (inadequate supplies, untrained workmen, skeptical manufacturers); he had also to overcome hostile editors, irate property owners, even novelists and poets! From its very inception in 1885 until long after its completion, the Tower was the center of a controversy so spirited as to appear almost incredible in retrospect. The design, appearance, cost, and safety of the structure became public issues. Prominent Frenchmen in all walks of life plunged into the discussion. Alexandre Dumas the younger and Guy de Maupassant were among the intelligentsia who signed a manifesto protesting the erection of the Tower; and the indignant poet Verlaine is said to have sworn never to visit that portion of Paris again. Newspapers took up editorial positions, *Le Figaro* going so far as to publish special issues on the subject. Several owners of property near the site instituted suit against Eiffel, insisting that the courts prohibit the construction of so dangerous a structure.

Whatever Eiffel's professional colleagues thought of his project, they very wisely refrained from signing manifestoes on the subject. Only the ornate Charles Garnier, architect of the Paris Opéra, circulated a petition to have it demolished by the government. The majority of professionals were at least respectful. Photographs taken during the construction indicate a stream of architects and engineers,

Fig. 129. Railroad Bridges at Busseau, 1864, and Garabit, France, 1880. Gustave Eiffel, engineer. In a series of increasingly daring steel bridges, Eiffel perfected the design and erection techniques used on the Tower.

Fig. 130. The Eiffel Tower, Paris, 1885–89. Gustave Eiffel, engineer; S. Sauvestre, associate. By means of his experience with the latticed pylons, Eiffel reached the unheard of height of 1000 ft.

top-hatted and frock-coated, being carried aloft on the dinky steam hoists and gravely clambering over the wooden scaffolding.

Gustave Eiffel fought his enemies to a standstill. When the French government, but half convinced of the soundness of his project, voted him only $292,000 of an estimated cost of well over a million dollars, Eiffel unhesitatingly supplied the balance out of his own pocket. He was confident, as he later told the Smithsonian Institution, that public opinion was on his side: "a crowd of unknown friends were ready to honor this bold enterprise as soon as it took form. The imagination of men was struck by its colossal dimensions."[9]

Eiffel's estimation of the temper of his age was correct. Once finished, the Tower immediately became, and has remained ever since, the most popular structure in France. In the single Exposition season of 1889 gate receipts at the Tower netted six-sevenths of the cost. It was described at great length in the world press, praised by Thomas Edison, who thanked God for "so great a structure," and recognized in the building field for what it was — a pacemaker for rigid structures.

Dramatic as was the public controversy surrounding the project, Eiffel's mastery of the technological dilemma confronting him was even more impressive. Already the *enfant terrible* of the European engineering fraternity, Eiffel had won an international reputation for his designs for the famous bridge at Garabit in France, the locks for the ill-fated French canal in Panama, and the supporting skeleton for Bartholdi's Statue of Liberty. All this was valuable preparation for the Tower, although in it he confronted problems of much greater magnitude than in these former projects. Indeed, he faced an absence of all those factors which the modern designer considers essential: a wide choice of specialized building materials, factories which could guarantee their prompt and regular delivery, suitable construction methods, trained workmen.

In evolving his general design, it is apparent that Eiffel drew upon his bridge-building experience, particularly that at Garabit (Fig. 129). He had come to "believe that it was possible to construct these [towers] without any great difficulty to a much greater height than

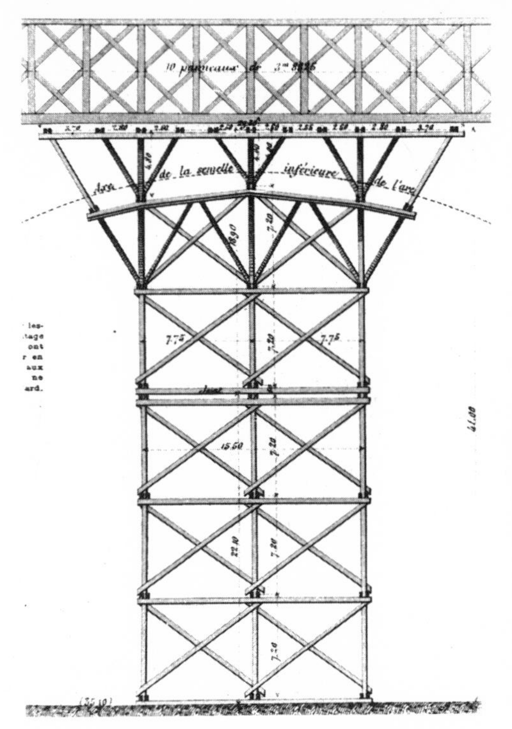

Fig. 131. Eiffel made a clear distinction between shop fabrication and on-site assembly. His office prepared complete drawings for all falsework and scaffolding (above). The Tower was also one of the first buildings to be fully photographed, in a planned fashion, as construction proceeded. Shown at right are a few of the many pictures made.

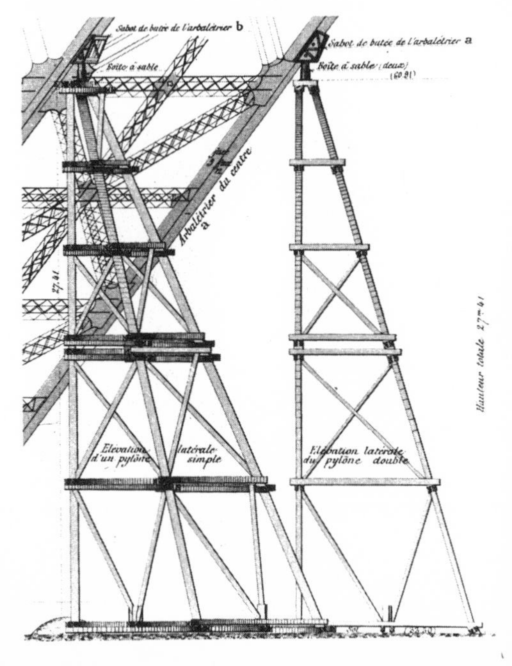

hitherto." The design was naturally based upon metal, since in such material constructions could now "be planned with such accuracy as to sanction the boldness which results from full knowledge." This boldness, however, did not blind him to caution: although the spread of the Bessemer process had by this time already made steel generally available, Eiffel conservatively chose wrought iron. He found its properties "remarkable, since it may be as readily employed in tension as in compression, and can be put together perfectly by riveting." This was not, of course, Eiffel's private discovery. In 1885, William LeBaron Jenney had already completed Chicago's first skyscraper, during which he would have met similar theoretical and practical problems. But the variation between American and French technology makes it unlikely that there was any direct exchange between Paris and Chicago.

Similarly, though the principle of reinforcing concrete with steel was already known, Eiffel stuck to stone masonry for his foundations, which rested in turn on concrete mats. And although he emphasized his faith in these foundations, he evidently realized that they were not ideal and cannily provided slots for eight-hundred-ton hydraulic jacks in each of the four piers.

Though Eiffel stuck to materials with which he was thoroughly familiar (he could not gamble on newfangled material like reinforced concrete, about which there was as yet no basic knowledge), he used them in an astonishingly modern manner. Satisfied though he pretended to be with stone masonry, his foundation design nevertheless anticipates contemporary reinforced concrete to a marked degree. Indeed, the four-inch wrought-iron bars which anchor superstructure to foundations also — "by means of iron clamps unite almost all parts of the masonry" — act as a rudimentary reinforcing. In much the same manner his detailing of the various wrought-iron members anticipates, in both profile and general shape, contemporary design of steel members (Fig. 131).

In the actual erection of the Tower, Eiffel was far ahead of the current European practice. Like Paxton, he made a clear distinction between shop and field operations, and had the entire seven thousand tons of ironwork completely fabricated at the factory, in-

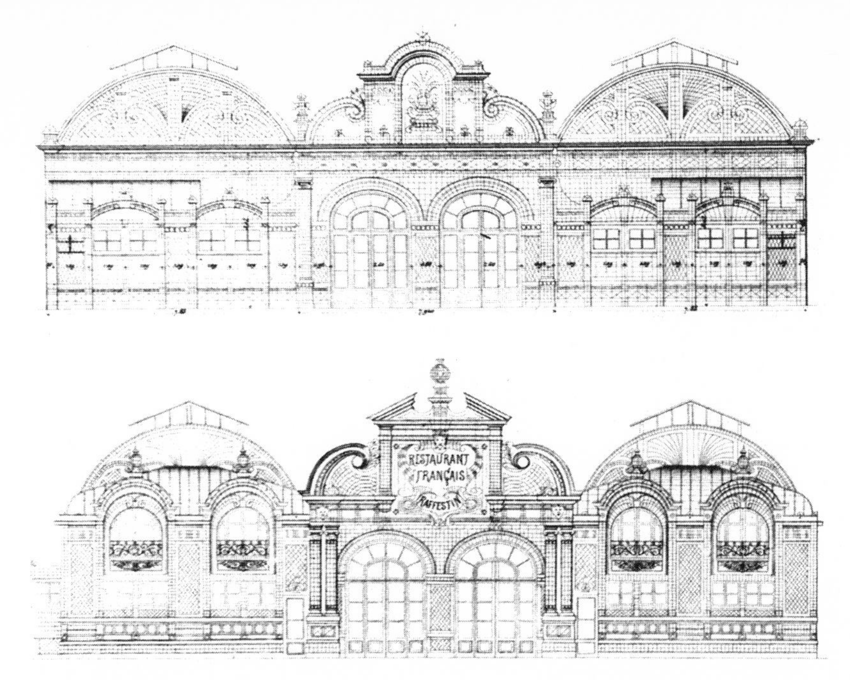

Fig. 132. Only in such "architectural" features as the Russian Restaurant (above) and French Restaurant (below) did the designers succumb to the prevalent eclectic idiom of fashionable architecture.

Fig. 133. The Eiffel Tower from the air, in one of the earliest examples of photography from a balloon. Note the "English-style" landscaping around the base.

cluding the punching of all holes and much of the riveting. This enabled him to use a relatively small crew (two hundred and fifty) of unskilled men at the site.

He designed his own scaffolding, making elaborate (and until then unheard of) provisions for the workmen's safety. This led one American magazine wonderingly to report to its audience: "It was feared that, unaccustomed to a very high scaffolding, few men could be found not subject to vertigo. But in [the construction of] the Tower they did not work high in the air with an open and dangerous footing. They were on platforms 41 feet wide and as calm as on the ground."[10] Eiffel was forced to design his own winches and rigging; and it is here that the technological lags against which he fought are most clearly apparent. By careful organization of each step in fabrication and erection processes, he was able to maintain control over the design, quality, and delivery-timing of his ironwork. But the industrial resources of his time simply did not permit a scaffolding and rigging system of similar efficiency. There was a time lag, between the design level of the Tower itself and that of its scaffolding and rigging, of at least half a century. This might have deterred other men: not so Gustave Eiffel. He forced the Tower through to completion without a single mishap, and thereby set the world new standards in the design and erection of very tall structures.

Not one of these three great designers was properly an architect, and only one of their three great structures can be properly classified as a building. Yet the popular acclaim which their works received shows how far removed from reality were the architects, the architecture, and the architectural critics of their day. There is little evidence that any of the three ever felt called upon to defend the appearance of their structures, much less to formulate any formal statement on personal esthetic standards. Yet even these three showed the esthetic ambivalence of their age when they ventured into those areas officially designated as "architecture": then their taste becomes faulty and insecure. This is most evident in the disparity between Paxton's glass buildings and his residential work: it seems incredible that the former could have so little influence on, or be so little influenced by, the latter. But the same unsureness may be detected in some of Roebling's detailing of the ornamental features of his pylons as well as in the nonstructural aspects of Eiffel's tower or in his Grande Gallerie for the Universal Exposition of 1878. Yet in their three masterpieces — the Palace, the Bridge, and the Tower — they show a lack of concern for current theories of the beautiful which is almost childlike in its absoluteness and purity. It is an ambivalence for which we can only be grateful. Their designs themselves carry the internal evidence of high discipline and precise selectivity — standards compared to which the turgid, flowery prose of professional critics like John Ruskin or James Jackson Jarvis appear tragically inadequate and ill-informed.

The successive impacts of these three structures upon American thought are readily traced in the periodicals of the day. Each as it appeared was added to the required itinerary of traveling Americans. The Bridge especially caught popular fancy: it was sold to yokels, woven into the music-hall routine, incorporated into our folklore. Of all sections of the population, the architects themselves seemed to have been least affected. The Crystal Palace was gracelessly mimicked in New York in 1853 and again at the Philadelphia Centennial in 1876. Roebling's bridge was quickly caricatured in the Queensborough and Manhattan spans which rose to flank it; and fragments reminiscent of Eiffel's ironwork were to be found embedded in later steel structures. Yet these were but distorted reflections of the original in which the clarity of the central concept was quite overlooked. It was as though no impulse, however strong or healthy, could remain long dominant in the ferment of post–Civil War America.

Each of the three structures was, in a way, the product of a golden moment of equilibrium, brief in time, special in character, delicate in balance. Their significance was dissipated before men of adequate stature could again appear to grasp it; and when Americans like H. H. Richardson and Louis Sullivan did appear, they came from a quite different background, equipped with a radically different perspective, so that they profited only indirectly from the explorations of the century's greatest structuralists.

6. 1860–1893
THE GREAT VICTORIANS

With the close of the Civil War, the rising industrialists achieved a smashing victory over the Southern slaveowners and their allies, the great merchants of the Northeast. The way was now clear for the full industrialization of the country, and the next forty years saw the incredibly rapid development of the entire continent. The defeat of the slaveowners removed the last barriers to the full economic development of modern capitalism. Never had the shift in class forces been more abrupt or profound; never had the change in upper-class ideology been more climactic; and never had its expression in architecture been more immediate.

The postwar period was lusty and inventive. At no time before or since has American building been so unselfconscious, so blithely forgetful, of the shadows of the past or the weight of the future. New tools, new materials, and new processes appeared with staggering rapidity to serve as new media for the builders. The metallurgical industries, enormously accelerated by the exigencies of war, were now moving into a position to supply structural steels, in unlimited supply and range of shapes and forms, to replace the less satisfactory wrought iron and cast iron of earlier days. Portland cement manufacture, begun by David Saylor at Coplay, Pennsylvania, in 1870, gave great impetus to brick and stone masonry. There was wide development in ceramics and clay products — necessary for fireproofing the new steel skeletons. Production of glass was industrialized, and the huge plate-glass windows of the Victorians were possible. Perhaps most indicative of all was what happened to wood. This had always been (and still is) America's favorite building material. It was abundant, cheap, and easily worked. In this material, with the advent of power-driven jigsaw and lathe, the esthetic aspirations of the period found their fullest expression. In maniacal enthusiasm it was cut, turned, twisted, tortured, and shaped: the medium par excellence for the symbols of the period.

All the buildings of the period gave characteristic expression to the narrow yet ardent interest of the American people in new machines and new processes. Even the official architecture — always a good barometer of upper-class taste — employed this lathe-and-jigsaw ideology, including even those buildings which pretended to be Gothic, Renaissance, or French. For this was the period of buoyant Victorian optimism when millowner and millworker alike saw only the promise (and not the problems) of the machine and mechanized production. It was only as the century drew to a close that "authentic" revivals began again to be demanded.

This period saw also the further separation of the architects from the main stream of American building. Under the severe strain placed upon it by expanding industrialism, the building industry itself began to undergo a qualitative change. No longer were building operations exclusively in the hands of small independents. In the rapidly growing cities speculative housing appeared; the mobility of labor steadily increased the ratio of rented to owned dwelling units, while even those who owned houses began more and more to buy them ready-built. The mills and factories were no longer the projects of scattered entrepreneurs, but huge plants for the new trusts. By the same token the building process itself changed. The independent artisans became skilled wageworkers; specialization set in, for it was no more possible to *build* the great factories with independent artisans than it was to *operate* them with cottage craftsmen. The building process began to be industrialized, and this was the main stream from which all advance in the future was to flow.

As a result of this same process, however, the big landlords — expanding in numbers, wealth, and complex administrative tasks — began to require the full-time professional services of the architect as designer and ideologue. In this shift, control of the design process

Fig. 134. Railroad trestle at Portage, N. Y. Silas Seymour, engineer. The need for economy and speed in railroad building precluded masonry viaducts. Fires like this one of May, 1875, forced engineers to turn from wood to steel.

Fig. 135. Grand Central Station, New York, N. Y., 1871. John B. Snook, architect. The railroad terminals needed unprecedented span. Here again speed and economy dictated skeletal vaults. The constant danger of fire led to the development of metal skeletons. This station was remodeled in 1889, replaced in 1903.

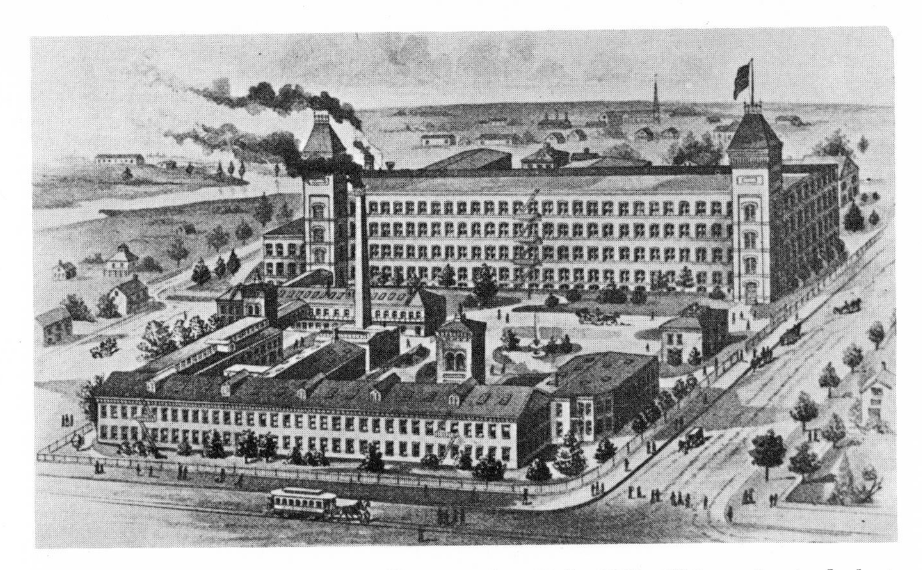

Fig. 136. Steam-powered cotton mill, Pawtucket, R. I., 1881. This mechanized plant employed heavy-timber construction, flat roofs. It was lighted by electricity, humidified and equipped with sprinkling system.

Fig. 137. Lucy Furnace, Pittsburgh, Pa., 1872. With abundant coal, iron and water supplies, Pennsylvania became the center of the new steel manufacturing industry. This Carnegie plant was a pacesetter.

passed out of the hands of the nameless artisan into those of the professional. How closely related this shift was to the triumph of industrialism over slavery is to be seen in these facts: the American Institute of Architects was founded in 1857; the first of fifty-odd architectural schools was founded at Massachusetts Institute of Technology in 1866; and the first architectural journal appeared in Philadelphia in 1868. Nor is it surprising that the concern of all these institutions should have been, from the first, "beauty." They were the organs of that group of specialists in the esthetic dimension of building whose inner content was always but a reflection of the ideology of the dominant sectors of society.

This phase of American building has been the source of endless confusion to the architect himself. The contrast between the relative simplicity and homogeneity of the main esthetic currents of American building up to 1860 and the turgid flood of the three decades which followed could only be ascribed to "an appalling decline in public taste," which marked "the lowest point to which American architecture had ever sunk." Since the consistent attempt was to explain esthetic phenomena in exclusively esthetic terms, the climactic change of the mid-century has been insoluble.

Actually, this confusion was the direct expression of profound social and technological change. Hitherto, three salient factors determined the character of American building: (1) The esthetic standards did not seriously conflict with the technological level of building. Neither the building materials (stone, brick, wood) nor the structural theories (post-and-lintel, load-bearing wall, arch-and-dome) differed in any important respect from those actually used by the Greeks and Romans. (2) The building types required by the economy were relatively few and simple, and could be readily fabricated with traditional materials along traditional lines. (3) The cleavage in society had not sufficiently sharpened to create wide divergencies in ideology — that is, to divide esthetic standards along class lines into "good" (upper-class) and "bad" (lower-class) taste. Industrialization had not yet succeeded in concentrating the production — and consequently the design control — in the hands of a small group of salaried specialists. Taste was thus still subject to some measure of democratic control.

Fig. 138. English High and Latin School, Boston, Mass., 1877. G. A. Clough, architect. With modern fireproof construction and a "German" plan, the new school replaced the earlier building where Louis Sullivan had studied before the fire of '72. It was the world's largest free public school.

Fig. 139. City Hall, Philadelphia, Pa. John McArthur, Jr., architect. Begun in 1872, but not completed until 1894, this emblem of civic pride became the symbol of municipal corruption.

In an indirect fashion, current concepts of the architecture of the nineteenth century are a characteristic expression of our misunderstanding of our own progressive past. For what the modern esthetician forgets, in smugly dismissing the period as "an all-time low in taste," is his own history. He is the product of precisely those institutions — the great architectural schools, the architectural press, the architectural profession itself — which our society perfected only in the latter part of the last century. In criticizing the "low" esthetic standards of his immediate predecessors and praising the "gracious, aristocratic" taste of the colonies, he is merely reflecting the contempt of many a self-made plutocrat for his own grimy origins and his yearning for the "gracious" days of colonial slavery.

THE IMPACT OF VICTORIAN TECHNOLOGY

But the fate of American building during this period was actually more in the hands of a constellation of remarkable outsiders than of the architects themselves: Morse with his telegraph, Bell with his telephone, Otis with his elevator, Goodyear with his gutta-percha, Westinghouse with his transformer and airbrake. These men were not necessarily the discoverers of the principles or inventors of the mechanisms they patented. In fact, Thomas Edison, the greatest of them all, invented nothing: generator, dynamo, transformer, cable, motor, and light bulb had all been proved possible by others. What he did accomplish, brilliantly, was the conversion of a laboratory curiosity into a socially productive reality. Nor was he exaggerating when he wrote that "everything is so new that each step is [a step] in the dark. I have to make the dynamos, the lamps, the conductors, and attend to a thousand details the world never hears of." It was the cumulative impact of the work of these men, in the years between 1865 and 1900, which completely transformed the character of both our buildings and the landscape in which they stood.

It was they who were to release building from its centuries-old limitations of size, density, and relationship. Thanks to them, the flow of men, ideas, things, both inside and between buildings, was speeded up to an extraordinary degree. Released from the necessity of face-to-face contact by the telephone and the pneumatic tube, the

Fig. 140. Proposed office building, New York, N. Y., c. 1870. Town and Davis, architects.

Figs. 141, 142. Grand Union Hotel, Saratoga Springs, N. Y., 1872 (left below) and Hotel Tampa Bay, Tampa, Fla., 1891 (right below), J. A. Wood, architect, were typical of the new type of resort hotel supported by the new urban wealthy and made possible by rapid transport.

scale of the department store could be almost indefinitely expanded. If freight and passengers could be moved vertically by elevators, then the sixth floor became as available as the first had formerly been. If by means of telephone, dictaphone, push button, and loudspeaker, a manager could direct five hundred workmen instead of fifty, then the factory could be that much larger. Or if — thanks to Mr. Edison's motors, wires, and current — one operator could control machines which had required ten men before, then the building automatically increased in complexity.

Contrariwise, the same new equipments tended toward decentralization. With electric power, the shop no longer had to be at the pithead. With the trolley, the worker no longer had to live in the shadow of the shop. And with bathroom and septic tank, steam heat and acetylene, the banker could comfortably escape with his children to the country. It needs emphasizing that in building, as elsewhere, "the machine" was not of itself responsible for the insane congestion of the cities. Implicit in it was equally the tendency toward decentralization. It was the character of our society, and not merely our tools, which determined the direction of nineteenth-century urbanism.

An increase in the mobility of people, things, ideas, not only made possible much larger and more complex buildings than had hitherto been possible; it also made possible a rapid specialization of building type. Jefferson had emphasized the distinction between building types in his home at Monticello as against the Capitol at Richmond. But his concern was, after all, a formal one — more a matter of ideological effect than technical effectiveness. It was enough if the one was recognizably the seat of a gentleman and the other the seat of a state power; and the truth was that neither operation would have been too seriously impaired had the buildings been switched. Matters were different a century later: many an architect might try to make his houses look like capitols but the violation of fact was difficult. For little by little all the Victorian gadgets had become organic parts of building. Indeed, heating and sanitation systems, light, power and communication systems, would come to be quite as important (and almost as costly) as the structural shell itself.

Fig. 143. A. T. Stewart's Department Store, New York, N. Y., 1860. First of the metropolitan department stores, it was the basis of a great mercantile fortune.

Fig. 144. Marshall Field & Co., Chicago, Ill., 1870. Daniel Burnham, architect. Its central rotunda with organ concerts has been a midwestern institution for decades.

This made American building a far more precise instrument of environmental control, but, by the same token, much less susceptible to undifferentiated use.

The high rate of urban expansion of the latter half of the century, plus the accelerating rate of technological advance, raised building obsolescence to new heights. In fact, an entirely new type of obsolescence appeared: *technological.* Hitherto a structure was seldom considered obsolete as long as it was physically sound. If a change in function was desired, say house into grogshop, a window was cut here, a counter built there, and a new sign hung to complete the metamorphosis. Of course, since change is uneven, many houses even today are converted into liquor stores. But the trend is entirely in the other direction.

The only important exception to this rule was in the domestic dwelling. Here technological obsolescence did not and does not operate as a law. As long as a house remained upright, it was considered by some landlord to be habitable; and, so perennial was the housing shortage, as long as a house remained upright, some tenant could be found to live in it. Except for local variations, there has been a housing shortage in America for at least a century. This is true both quantitatively and qualitatively, and all the mad and avaricious expansion of the Victorians scarcely changed the picture. Nevertheless, even in the house, the optimum standard was rising so that, by the end of the century, a house without central heating, bath, and lights was definitely substandard, even though the vast majority lay in precisely that category.

Today it is apparent that the origins of our most spectacular architectural accomplishments are bedded in the rich and maculate Victorian era. Those gentry clothed our cities in such confusion, waste, and appalling ugliness that it has been difficult for many critics to "see any good" — indeed, to see anything at all — in their work. More than refined esthetic standards are necessary to such an exploration, for their task was gargantuan and complex, and there was no time to be finicky. Those very scientists, inventors, and builders whose grimy work seemed so remote to the stylish *fin de siècle* architect were in fact making great contributions to contempo-

Fig. 145. “A Hot Night in the Tenements.” Conditions in the congested Eastern seaboard slums had become local scandals by the 1880’s.

Fig. 146. Model Tenements, Brooklyn, N. Y., 1876–79. Field & Son, architects. Built by the philanthropist Alfred T. White, they were an early effort to provide sanitary, low-cost, non-profit housing.

rary building — greater perhaps than all but a handful of the architects themselves.

BELLAMY, VEBLEN AND "THE HARSH DIVORCE"

It would be inaccurate to say that these peripheral developments were ignored by the building field. Despite its general backwardness, it adopted some of them with astonishing rapidity. (For example, the elevator and the skyscraper were Siamese twins, largely perfected in a short two or three decades.) But the resistance among the architects themselves was immense. They romantically rejected the material promise of mid-Victorian life or they accepted it piecemeal. In neither case did they show any grasp of its potentials. Indeed, the only coherent perspective of the period was not to be found among the professionals but in the utopian tract of Edward Bellamy, *Looking Backward*.[1] In this remarkable book, Bellamy displayed an uncanny anticipation of the future development of technology. His utopian Boston had a twenty-four-hour broadcasting system and every room was wired for sound. Chimneys — in fact, combustion itself with its soot and smoke — had completely disappeared: electric power was everywhere, cheap, abundant, clean. Artificial lighting was indirect, air conditioning was commonplace, housework was mechanized. Yet this was the least distinctive aspect of Bellamy's perspective. What lifted him above the ruck of mid-Victorian idealism was his naïve yet luminous vision of how technological advance was to deepen and enrich the whole course of American life.

The Boston he saw from his rooftop, that first summer morning in the year 2000, had been completely rebuilt into a modern garden city:

> Broad streets, shaded by trees and lined with fine buildings, for the most part not in continuous blocks but set in larger or smaller enclosures. . . . Every quarter contained large open squares filled with trees, among which statues glistened and fountains flashed . . . the public buildings [were] of a colossal size and an architectural grandeur unparalleled.[2]

The entire distributive apparatus of the city had been completely

Fig. 147. Central Park, New York, N. Y., 1859–1879. Frederick Law Olmsted and Calvert Vaux, landscape architects. This bird's-eye view from the southeast corner shows its remarkable scale and great variety.

Fig. 148. Michigan Avenue and the Lake Front, Chicago, Ill. This 1889 view shows Adler and Sullivan's new hotel.

revamped, so that each ward was serviced by a single branch department store. Each of these was connected by pneumatic tube to a huge central warehouse from which all deliveries were made, also by pneumatic tube. This warehouse was the apotheosis of American rationalization — "a giant hopper into which goods are being constantly poured by the trainload and shipload, to issue at the other end in packages of pounds and ounces, yards and inches, pints and gallons." Yet the mechanism was only incidental to the social order it served. Satisfaction and comfort were paramount considerations. Hence the architectural design of a typical store:

> a vast hall full of light, received not alone from the windows on all sides but from the dome, the point of which was a hundred feet above. Beneath it . . . a magnificent fountain played, cooling the atmosphere to a delicious freshness . . . walls and ceilings were frescoed in mellow tints to soften without absorbing the light. . . . Around the fountain was a space occupied with chairs and sofas.[3]

In Bellamy's Boston labor-saving machinery had been used "prodigiously," but the emphasis was upon the elimination of drudgery rather than in the machine *per se*. Ridding the houses of the dust-catching bric-a-brac of General Grant's epoch had not only reduced woman's work: it had increased her social stature. Central kitchens and dining rooms not only took the housewife away from the stove: they gave her the time for a career of her own if she so desired. If there was an absence of detail in Bellamy's architecture, it was because he was interested in the social function of building, not its street façade. Victorian society was drowning in detail, rich and proliferating. What it needed was organization. Any engineer could install his broadcasting system: the real problem was to see that the quality of the broadcast programs themselves was good and that all had access to them.

Bellamy managed to project a picture of urban life which combined New England Utopianism with the characteristic optimism of the early Victorians. To it, however, he added a concept of organization — whether social or architectural — which was unprecedented in breadth and completeness. He saw what could be done with the materials at hand where his contemporaries in the building

Fig. 149. Southwest view (above) and interior (below), St. John the Divine, New York, N. Y. Result of one of the nation's first competitions, this prize-winning design by Heins and LaFarge was ill-fated from the start. Begun in 1892, turned over to Cram and Ferguson in 1911, it remains unfinished to date.

field, bemused by the sheer multiplicity of materials, could not so much as visualize a coherent city block.

The split between ideologue and technician developed most sharply in the Northeast, as a result of the industrial expansion immediately preceding and following the Civil War. By the nineties this contradiction was already well developed — was, in fact, a principal topic of discussion among the architects themselves. Thus, in 1892, A. D. F. Hamlin ruefully observed that:

> Engineering had monopolized whatever real progress was being made in building. Metal construction was coming into general use for bridges and for structures with large roofs, such as railroad stations and exhibition buildings — the most characteristic products of the constructive skill of the time. These works were entrusted to engineers; the architects were so preoccupied with their mistaken efforts to resuscitate historic styles that they wholly failed to discover the possibilities of the new material, and scornfully abandoned it.[4]

And another prominent critic of the day, Montgomery Schuyler, saw the situation as critical:

> The architect resents the engineer as a barbarian; the engineer makes light of the architect as a dilettante. It is difficult to deny that each is largely in the right. The artistic insensibility of the modern engineer is not more fatal to architectural progress than the artistic irrelevancy of the modern architect. In general, engineering is at least progressive, while architecture is at most stationary. And, indeed, it may be questioned whether without a thought of art and, as it were, in spite of himself, the engineer has not produced the most impressive as certainly he has produced the most characteristic monuments of our time. . . . What may we not hope from the union of modern engineering with modern architecture, when the two callings, so harshly divorced, are again united![5]

However clearly Schuyler saw the dichotomy in nineteenth-century building, he remained the prisoner of his time and class. He could not see that the harsh divorce between architect and engineer was the inevitable product of the social order. On the one hand, industrial production demanded better performance of building; consequently, the industrialists were willing to subsidize progress in engineering and technology. But, on the other hand, the ideology of these same men was becoming increasingly conservative. They re-

Fig. 150. Biltmore, near Asheville, N. C., 1895. Richard Morris Hunt, architect. This prefabricated estate including dairy, stables and private railway stop was carved out of the Appalachians in three years.

Fig. 151. Potter Palmer house, Chicago, Ill., 1882. Cobb and Frost, architects.

quired of the architecture which they subsidized — especially that architecture with any pretense to symbolic or monumental significance — that it make explicit this conservative world view.

It is truly ironical that neither these capitalists nor their architects could conceive of, much less create, any new iconography of power. All they could do was to borrow those of the discredited past — the feudal barony, the Roman empire, the Egyptian theocracy. William H. Vanderbilt might protest to his architect that he and his family were "plain, quiet, unostentatious people" whose tomb needed to be only "roomy and solid and rich." But the house which that architect built for his son on Fifth Avenue was entered through copies (reduced in scale) of the doors which Ghiberti had done for the Baptistry in Florence. Mrs. Potter Palmer was ensconced on Chicago's North Shore behind crenellations and battlements borrowed inexpertly from Norman England. The new San Francisco tycoons, less anachronistic if no less authoritarian, were adopting the imperial manner of Napoleon III and Eugénie. Everywhere, the rich and powerful were compressing the functional content of architecture into formal configurations antithetical to it. The dichotomy was acute. Nor would it be resolved until esthetic and technical standards could evolve in the same direction and at the same rate of speed.

This reactionary application of eclecticism, so different from what the radical Romanticists of the forepart of the century had envisioned, ultimately smothered all traces of originality in the architecture of the East. Recent studies by Vincent Scully[6] have established the fact that an undercurrent of Yankee sanity survived on into the eighties. Christened by him the Shingle Style, this movement was centered in Boston and found expression almost exclusively in a series of handsome, open, and hospitable houses for the Boston suburbs and the New England seacoast. It produced such paradigms of domestic felicity as H. H. Richardson's house for the Stoughtons and the William Low house by McKim, Mead and White (Figs. 152, 153). It would be rewarding to analyze in some detail the social and cultural milieu of the people who built these houses. Clearly, in their confidence and lack of pretension, they

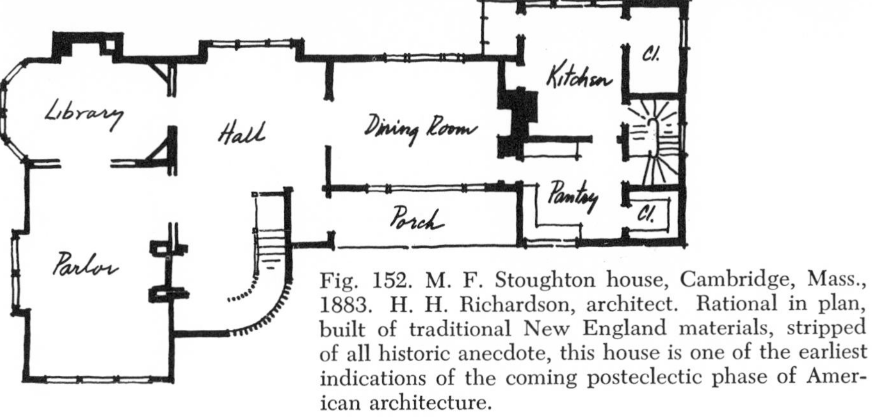

Fig. 152. M. F. Stoughton house, Cambridge, Mass., 1883. H. H. Richardson, architect. Rational in plan, built of traditional New England materials, stripped of all historic anecdote, this house is one of the earliest indications of the coming posteclectic phase of American architecture.

Fig. 153. W. G. Low house, Bristol, R. I., 1887. McKim, Mead and White, architects. For a few years, before they joined the Neoclassic establishment, these young protégés of Richardson used his idiom.

represented standards quite different from the world of high fashion and great wealth. And that perhaps explains why, for all its vigor and promise, the Shingle Style had so little direct influence on the course of events. Through Richardson himself, the idiom was transplanted to Chicago, where it clearly became a factor in Frank Lloyd Wright's "prairie house." But Richardson was dead by 1886; and McKim, Mead and White, by 1890, were well on the way to becoming the most imperial of all *fin de siècle* eclectics.

In the East, no resolution of the harsh divorce was possible. On the contrary, the gap between technology and ideology was steadily to deepen as the century closed. And as the gulf widened, the iconography got steadily worse. There was some protest from outside the field. The novelist F. Marion Crawford, son of the sculptor-friend of Greenough, attacked "false taste" in art and called for a return to the "simplicity and grace" of eighteenth-century architecture. The critic Clarence Cook, in a biting essay in 1882,[7] declared that "where architects abound, the art of building always deteriorates. . . . It is to the architect that we owe the ugly building[s] that offend us." Richard Morris Hunt especially offended him: "He has certainly loaded the earth with some of the most ungainly . . . structures that make our streets such a misery." But these genteel cries of protest went unheeded if not unheard. The sardonic Thorstein Veblen, observing the antics of the ruling classes in 1899 with caustic perspicacity, saw that nothing could escape the corroding effects of their new esthetic standards of "conspicuous waste and predatory exploit."

> It would be extremely difficult to find a modern civilized residence or public building which can claim anything better than relative inoffensiveness in the eyes of anyone who will dissociate the elements of beauty from those of honorific waste. The endless variety of fronts presented by the better class of tenements and apartment houses in our cities is an endless variety of architectural distress and of suggestions of expensive discomfort. Considered as objects of beauty, the dead walls of the sides and back of these structures, left untouched by the hands of the artist, are commonly the best feature of the building.[8]

Aside from their homes, Veblen said, the "imperious requirements" of these bigwigs could best be observed in the architectural style they imposed on the orphanages and old ladies' homes which they were then endowing. Here

an appreciable share of the funds is spent in the construction of an edifice faced with some aesthetically objectionable but expensive stone, covered with grotesque and incongruous details, and designed — in its battlemented walls and turrets and its massive portals and strategic approaches — to suggest certain barbaric methods of warfare. The interior of the structure shows the same pervasive guidance of the canons of conspicuous waste and predatory exploit![9]

FALSE SPRING IN CHICAGO

Out in the Middle West a somewhat special set of circumstances was to lead to different results. Chicago, already the metropolis of the inland empire, had been all but destroyed in the Great Fire of October 8, 1871. In the decades which followed it was the scene of unprecedented expansion. This construction alone would have furnished the material basis for a large and prosperous group of architects and engineers. But the rise of the famous "Chicago School" of architects — and its unchallenged ascendancy in the nation throughout the next several decades — cannot be wholly explained in terms of large and numerous commissions. These men developed in the electric environment of a vigorous and progressive capital, intellectual center of the Midwest. In the older cities of the eastern seaboard, the Wall Street capitalists had already consolidated their power and settled down for a long period of intensifying conservatism. Chicago, on the other hand, was the pivot of industrial and agrarian forces which resisted, for the time at least, the long arm of eastern monopoly. It faced huge new problems and had no choice but to try new solutions. Internally, it was witnessing a rapid polarization of social forces marked on the one hand by the rise of the grain, packing-house, railroad, and manufacturing industries, and, on the other, by the great trade-union movements. It was thus the home both of robber barons and Haymarket martyrs, of the Pullman massacre and the eight-hour day. And beyond the city proper lay the great democratic hinterland, with its agrarian Populism.

It was this added dimension which distinguished the atmosphere of the city from that of the East, and subtly but positively affected the opinions and the work of its building designers. It must thus not be considered accidental that the great trilogy of American architects — Richardson, Sullivan, and Wright — is permanently identified with Chicago. Elsewhere they may have worked, but nowhere, during that golden era of three decades, did they find a more hospitable environment. They were not by any means the only fine architects in Chicago at that time — merely the greatest. With them at various times were associated Adler, Burnham, Root, and Holabird, and others — all of them loosely affiliated in the Chicago School. Whatever differences there may have been among them, there were many points in common. And these points were best exemplified in the three leaders.

Henry Hobson Richardson entered L'Ecole des Beaux Arts in Paris in 1860; and when, five years later, he disembarked in Boston, he carried with him a magnificent instrument — one that was to make him the first great architect since the days of Mills and Latrobe: the clarity and precision of French academic thought at its best. The esthetic character of his work and its impact on contemporary American building have been amply treated, notably in Professor Hitchcock's book,[10] so that it requires little attention here except to venture an opinion that too much has perhaps been made of this aspect of his work. For Richardson's largest significance lies not so much in the architectural vernacular he perfected as in his role of restoring the social function of the architect to something of its former prestige.

His work in the East had already achieved the high praise of imitation when he came to Chicago in 1885 to build the famous wholesale store for Marshall Field (Fig. 154). It created an immediate sensation. But the thing about the Field Building which Louis Sullivan was quick to appreciate was not so much the majestic scale of arched bays and rusticated stonework as the simplicity and clarity of its organization. Here he saw a "direct, large, and simple" mind at work — the first architectural mind in half a century which showed control of the medium in which it worked. Richardson's

Fig. 154. Marshall Field Wholesale Store, Chicago, Ill., 1885–87. H. H. Richardson, architect. The publication of this design prompted Sullivan to revise his Auditorium Building (see Figs. 158–59).

Fig. 155. Allegheny Jail and Court House, Pittsburgh, Pa., 1884–87. H. H. Richardson, architect.

work enjoyed then and later a great reputation for its "composition." But what was this composition if not an expression of the skill and understanding of the designer, a knowledge of the forces with which he was dealing, an ability to analyze their contradictions and then resolve them into a workable solution? If the churches, store buildings, and railroad stations which Richardson was then designing impressed the public with their outward appearance of order, organization, and plain good sense, it was because Richardson so largely understood the complex organisms of modern capitalism — the huge department store, the large and worldly metropolitan church, the modern library — he was called upon to house. He was like Roebling and Eiffel in this respect — capable of grasping the large outlines of the problem before him, of analyzing the forces involved, organizing them into a coherent and workable plan.

Henry Richardson died at the age of forty-eight, when his career was at its height, and one can only speculate as to what he might ultimately have achieved. It is possible that he would not have gone much farther, for his chief contribution lies in the field of plan rather than structure. He seems to have had no deep understanding of structural design or interest in constructural process — that is, in *building.* He used current technical advances — elevators, steel frames, electricity — competently but casually. The rheostat he put into the Glessners' new mansion for dimming the dining-room lights must have caused a sensation among the guests at the housewarming; otherwise the house is technically unexceptional, differing only from other mansions in its plan — that is, its spatial organization — and hence in its form (Fig. 157). Richardson was not of a speculative turn of mind, had no philosophical pretensions, and did not commit himself to paper; so we can take Sullivan's estimation of him ("direct, large, and simple") as adequate. Yet his invasion of the Chicago scene was destined to have tremendous repercussions, leading directly to the century's most important development — the appearance of the Chicago School.

Stylistically, his work remained till the end eclectic. From Trinity

Fig. 156. Trinity Church, Boston, Mass., 1873–77. H. H. Richardson, architect. The first mature demonstration of Richardson's talent, this prize-winning design was the start of his fame.

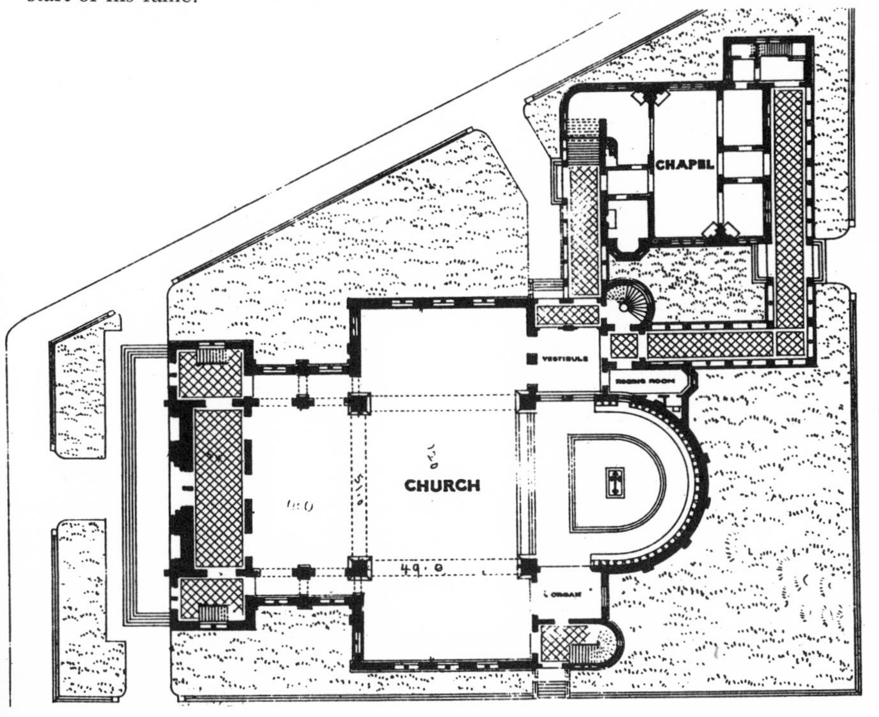

Church in Boston, through the Pittsburgh Court House to the Glessner house and Marshall Field Building in Chicago, he employs load-bearing masonry modeled after Romanesque precedent. There is, however, a steady process of simplification at work. Explicitly historic ornament is little by little abandoned. And the picturesque composition of which he was such a master is gradually subordinated to a system of masses and voids which grows directly from extremely rational plans. In this sense, eclecticism had served Richardson in exactly the way that Robert Dale Owen had urged long before — i.e., as a liberating force from the "Procrustean Classic."

The irony was to be that Richardson had no legitimate heirs in the East. His building was widely copied, his work never really studied. It was always the elevation, never the plan, that stylish architects were intent on copying.

But upon one architect, at least, Richardson's work had a fertilizing effect: Louis Sullivan. If one compares the Field Building with the final design for the Auditorium Building, the direct relationship is immediately apparent. This impact of Richardson upon Sullivan is readily apparent if one studies the first designs Sullivan did for the Auditorium project. Here, although the basic plan had been developed, Sullivan had been playing around with trivial façades — cute Swiss balconies, French Renaissance towers, typical Victorian mansards. These served to obscure rather than to clarify the basic organization of the huge project. Sullivan was well aware of this dilemma but unable to escape it — until he saw the Field Building. Here were façades which precisely expressed the internal organization of the project. Eighty years later, this does not sound like an especially radical discovery, but the young Sullivan saw in it an immensely important principle, the lack of which had vitiated building for almost a century. He was quick to apply it in the Auditorium Building and to develop it theoretically into his famous axiom: "Form follows function."

This organizing principle was the major contribution which Richardson made to architecture. It found its most fertile ground in Chicago. Here, and almost only here, the preconditions were right for further growth.

Fig. 157. John J. Glessner house, Chicago, Ill., 1885. H. H. Richardson, architect. Like the Stoughton house, a handsome, well-built, self-confident example of Richardson's work for the urban well-to-do.

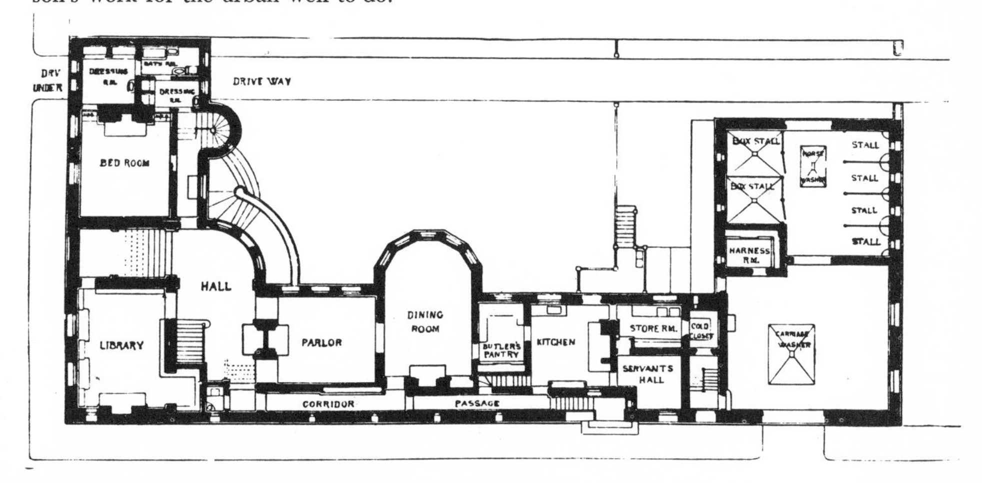

SULLIVAN, GIANT OF THE MIDWEST

Three great cities shaped Louis Sullivan's life — Boston, Paris, Chicago — and the imprint of all three is clearly evident in his buildings and in his thought. From the Boston of his birth came the warm humanism of the transcendentalists and the Whitmanesque passion for democracy; from his extended stay in Paris came intellectual discipline, ability to study, and a profound interest in science. But from raw, expansive Chicago came the inspiration and the opportunity for the most creative years of his life. It is to Chicago that he belongs; and he sensed this fact the first time he visited Chicago as an ambitious and confident youth of seventeen. It was still largely in ashes from the Great Fire of 1871 and rocking from the current Panic of 1873; yet years later, he still remembered it as

> magnificent and wild: a crude extravaganza, an intoxicating rawness, a sense of big things to be done. For "big" was the word . . . and "biggest in the world" was the braggart phrase on every tongue . . . [the men of Chicago] were the crudest, rawest, most savagely ambitious dreamers and would-be doers in the world . . . but these men had vision. What they saw was real.[11]

Sullivan went to Paris to study at the Beaux Arts but his early fascination for the precision of French scholarship had gradually given way to a healthy fear of the dry rot of French academism. Straight as a homing pigeon he returned to Chicago. In 1879 he went to work for Dankmar Adler and on the first day of May, 1881, became the partner of that most able man. Thus began the meteoric career of Louis Sullivan. From the first, he was destined to occupy an anomalous position in the bustling, competitive world of Chicago architecture. Like his contemporaries — Root, Burnham, Richardson — he was a successful architect. Indeed, in the Auditorium Building he and his partner, Dankmar Adler, walked away with the largest and most costly project of the times. Unlike some of his contemporaries, he was successful in a much more profound sense, since his best buildings unquestionably rank at and near the top in American building. He was a great designer in the same fashion that Richardson was great: in his ability to analyze a complex problem, master it, solve it finally in a brilliantly articulated design. But he was more than that, as he insisted any architect worth his salt

Fig. 158. The Auditorium Hotel, Chicago, Ill. Adler and Sullivan, architects. One of several early studies, a coherent plan and sensible structure were still masked by a mediocre, eclectic facade.

Fig. 159. The same project as redesigned by Sullivan after he had studied the published drawings of the Marshall Field Store (Fig. 154).

should be: he was also a theoretician, an ideologue, a philosopher.

If Sullivan's chief impact upon American building was to be through his actual projects rather than his written word, this was due not so much to faulty prose as to a peculiarity of the architectural profession: namely, that architectural concepts are historically visual. In the truest sense of the word, the architect has never been literate; to him a picture is worth a thousand words, and since the introduction of photography, ten thousand. Thus a hundred architects saw pictures of Sullivan's Wainright Building for one architect who read his articles or attended his lectures; and it is to these structures that we first must turn in analyzing Sullivan's effect upon American building.

Of what did this work consist? One hundred and twenty-four structures — whose dissimilarity in appearance, disparity in quality, and range of types is striking — are attributed to Sullivan.[12] In retrospect, this list seems contradictory. It is difficult at first to believe that the Auditorium and the Schlesinger Building were designed by the same man; that any design so mediocre as the Borden residence could have preceded the Babson house; or that one man could with equal ability master the plan complexities of the Auditorium and the ultimate simplicity of the Getty Tomb. Despite this, there is an inner continuity to the man's work. No architect ever had less of a vested interest in a "private" style than Louis Sullivan. He was quick to see the advantages of other men's work, and quick to adapt them to his own uses. Thus his profound respect for Richardson's Field Building was quickly reflected in his final designs for the Auditorium. Yet Sullivan was in no sense a plagiarist. He was a man in search of an architectural idiom which would be, not his own personal property, but that of the American people. It must express, not his culture, but that of his fellow countrymen. It was merely his task, as he put it, to catch and make incandescent that culture, "as the poet, looking below the surface of life, sees the best that is in the people." To this end, both in his buildings and his writings, he dedicated the major part of his energies. And it was only as adversity, ill-health, and isolation overtook him that his search for an appropriate idiom was reduced to the subjective metaphysics of a personal style of ornamentation.

Fig. 160 (left). Wainwright Building, St. Louis, Mo., 1891. Adler and Sullivan, architects. First of the "classic" skyscraper formulations of base, shaft, cap.

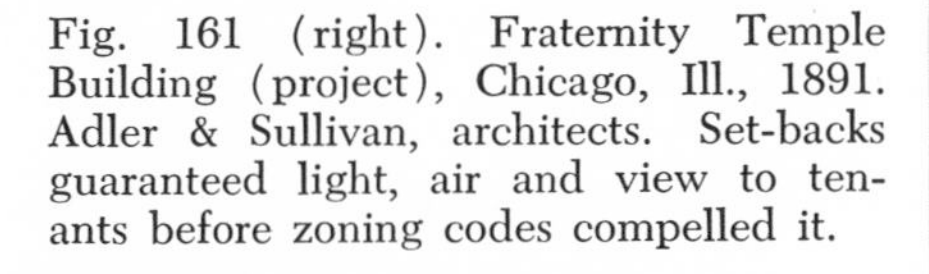

Fig. 161 (right). Fraternity Temple Building (project), Chicago, Ill., 1891. Adler & Sullivan, architects. Set-backs guaranteed light, air and view to tenants before zoning codes compelled it.

Fig. 162 (left). Guaranty Building, Buffalo, N. Y., 1894–95. Adler & Sullivan, architects. Doubled piers and recessed spandrels to accentuate the "proud and soaring" lines of the skyscraper shaft.

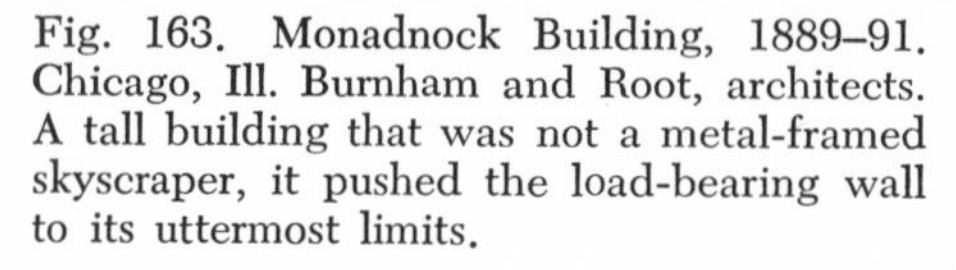

Fig. 163. Monadnock Building, 1889–91. Chicago, Ill. Burnham and Root, architects. A tall building that was not a metal-framed skyscraper, it pushed the load-bearing wall to its uttermost limits.

Fig. 164. Reliance Building, Chicago, Ill., 1890–94. Burnham and Root, architects. An elegant, metal-framed skyscraper by the same firm, this little tower seems astonishingly up-to-date, 70 years later.

In building design proper, his work showed a steady progression, but with equally climactic breaks. Five buildings marked this progress: the Auditorium, the Wainwright office building in St. Louis, the Guaranty office building in Buffalo, the Schlesinger and Meyer store, and the National Farmers' Bank in Owatonna, Minnesota. In addition to his mastery of plan, Sullivan demonstrated a growing mastery of structure. The Auditorium was of hybrid construction with load-bearing exterior walls; its principal innovations were the foundations and interior framing of cast and wrought-iron members.[13] Here the multi-storied, many-windowed structure is carried as far as it could go without the complete steel frame. Burnham and Root's splendid Monadnock Building of four years later merely restates this fact in a style more pleasing to contemporary taste. In the almost barbaric splendor of the public rooms of the hotel and theater, Sullivan used ornament, textures, and a palette of rich color which upper-class Chicago of 1887 found well suited to its needs.

Sullivan invented neither the steel frame nor the multistoried building; but in four years, in the Wainwright and Guaranty buildings he produced the prototype, which in plan and façade was not to be materially improved upon for half a century (Fig. 162). Sullivan was much impressed with the esthetic potentials of tall buildings at this time and sought, by careful organization of every detail of his exteriors, to make the maximum use of verticality. In the Guaranty Trust in Buffalo, he carried this particular formula to a superb conclusion. Yet in 1893 he must already have begun to question the validity of this generalization and wonder how far it could be extended, for in that year he designed the Meyer Building. Here, for perhaps the first time in modern architecture, the deliberate emphasis was placed upon horizontality.

Now, the whole discussion of whether a skyscraper should be "vertical" or "horizontal" was in fact academic; structurally, the steel frame was a static cage. Floor loads are horizontally conveyed to vertical columns and thus transmitted to the ground. And when, in 1899, Louis Sullivan came to what was to prove his most pregnant design — that of the Schlesinger and Meyer store — he recognized this fact. Neither verticality nor horizontality is emphasized. Since maximum light was a prime consideration, he reduced both piers and spandrels to the absolute minimum; and then, as if to emphasize the static nature of the skeleton, he surrounded each window with a fine but very positive frame, so that the honeycomb character is inescapable (Fig. 165).

It is in this remarkable building too that we find one of Sullivan's most polished statements on the subject of ornament. The first two floors are faced by a bronze-and-plate-glass screen, structurally and esthetically independent of the main building. The ornament is fantastically detailed and fine, so that much of it actually serves as a textural surface. All of it is typically Sullivanesque — sculpture done with a 4–H pencil, a curious *mélange* of geometry, Art Nouveau, and Classicism. But it is notable on several counts. The entire screen is designed for mass production in bronze. As a whole, it is perhaps the first modern shop front in the country; its design is highly integrated, providing so well for hardware, awnings, ventilators, illumination, that five decades of continual use have required no alterations.

This screen has been criticized as being out of keeping with the remainder of the façade. And the dichotomy between the two is indeed acute (bottom, facing page); it is hard to believe that two such divergent patterns are from the same hand. But they do clearly express the two major themes of Sullivan's development, and they anticipate the last phase of his career — that series of small-town banks which he was to do in the Midwest. In these buildings he was, in a certain sense, a free man. Though he did not recognize it for many years, his big-city career was ended in 1899, with this department store. And only when he came to write his *Autobiography* was he able to see, from the perspective of thirty years, that the death sentence of his hopes had been handed down at the Exposition of 1893.

The lifework of Louis Sullivan lies in two great deposits — his buildings and his writings. The two are, so to speak, complementary rather than repetitive: what he could not achieve as a designer of actual buildings he sought to establish in the field of polemics. They mirror the contradiction which ultimately destroyed him, although

Fig. 165. Schlesinger & Meyer Department Store, Chicago, Ill., 1899. Louis Sullivan, architect. Here the static, non-directional structural system and volumetric honeycomb of the skyscraper are given direct formal expression. In contrast, the bronze screen at street level, with its non-structural flamboyant plasticity, looks like the work of another man: a typical Sullivanesque paradox.

his accomplishments in both fields make him unquestionably one of the greatest figures in modern architecture. In the design of great and complex buildings he was without peer; his analysis of the social function of building and the special role of the architect must — for its clarity, courage, and perception — remain the classic on the subject. There may be appalling lapses in both areas — scattered buildings of poor design or pages of florid turgidity; but these are not central to the issue, merely the excrescences of a fabulously fertile and diversified individual.

Although he proved himself time and again to be the most competent architect of his period — proved it in an astonishing variety of buildings from tombs to department stores — Sullivan had singularly little to say about the practical aspects of building. Perhaps it was because his executed designs so adequately expressed his opinions in steel and concrete; there was little that needed to be added in words. In order to design a satisfactory auditorium, one had to comprehend acoustics; to turn out a satisfactory steel-framed skyscraper, one had to understand structure. This was self-evident; and decades before the rest of the profession, Sullivan had furnished the prototype and the program for the tall office building. He went much farther; though he personally was stirred by the esthetic possibilities of a tall building ("the force and power of altitude . . . a proud and soaring thing"), he was nevertheless aware of other and more important aspects. In his second tall office building (Schiller, 1891), he was already sufficiently troubled by the problem of light and air to give up a large portion of the plot to light wells. In his design for the unrealized Fraternity Temple (1891), he completely anticipated the free-standing cruciform tower, with diminishing setbacks, which the architects and promoters of the thirties were to glorify. Ultimately he came to the conclusion that the skyscraper carried the seeds of its own destruction.

> The tall steel-frame structure may have its aspects of beneficence; but so long as man may say: "I shall do as I please with my own," it presents opposite aspects of social menace and danger . . . the tall office building loses its validity when the surroundings are uncongenial to its nature; and when such buildings are crowded together upon narrow streets or lanes they become mutually destructive.[14]

Fig. 166. Getty Tomb, Graceland Cemetery, Chicago, Ill., 1890. Adler and Sullivan, architects. Solid masonry affords a perfect plastic medium for Sullivan's almost calligraphic ornamental system.

Thus, Louis Sullivan was driven always, by the logic of his own definition of an architect, from the field of practical technique to social, political, and philosophic speculation. Owing to his own deep scientific interests, his training in France, most of all to his close and fruitful association with Adler, Sullivan could always master the most advanced technological developments of his day. With Adler there, no problem in foundations, acoustics, electric illumination, or ventilation would long go unsolved. In this respect, Sullivan's best work resembles that of Eiffel or Roebling; but unlike the great engineers, he was unable to regard his structures as isolates. It was not enough that a building perform efficiently and economically; it must in addition *express* that function. As eloquently as Jefferson and Greenough, and even more explicitly, Sullivan fought to make architecture a tool of social and political progress.

He told the Chicago Architectural Club in 1899:

> Accept my assurance that [the architect] is and imperatively shall be an interpreter of the national life of his time . . . you are called upon, not to betray, but to express the life of your own day and generation . . . a fraudulent and surreptitious use of historical documents, however suavely presented, however cleverly plagiarized, however neatly repacked, however shrewdly intrigued, will constitute and will be held to be a betrayal of trust.

The real function of the architect, he never tired of insisting, was:

> to vitalize building materials, to animate them with a subjective significance and value, to make them visible parts of the social fabric, to infuse into them the true life of the people, to impart to them the best that is in the people, as the eye of the poet, looking below the surface of life, sees the best that is in the people.[15]

If one really wanted to know the causes for the standards of taste then prevailing in American architecture, one must look deep. For, he said:

> Architecture is not merely an art, more or less well or more or less badly done; it is a social manifestation. If we would know why certain things are as they are in our architecture, we must look to the people; for our buildings as a whole are an image of our people as a whole, although specifically they are the individual images of those to whom, as a class, the public has delegated and entrusted its power to build. Therefore by this light, the critical study of architecture becomes . . . in reality, a study of the social conditions producing it.[16]

A critical study of the conditions producing Chicago's architecture should have revealed to Sullivan precisely why so much of it was bad. He had as clients many of those very men "to whom the public had delegated . . . its power to build." And all his abilities as a designer, all his powers of persuasion as speaker and pamphleteer, could not bridge the sharpening gap between the aspirations of the mighty and the needs of the people as a whole. For a short period — roughly from the Great Fire to the Columbian Exposition — Chicago had hung in a sort of miraculous equipoise, battleground for Midwestern equalitarianism and rapacious monopoly. And during this period Sullivan had as patrons and friends some of Chicago's richest men: they had subsidized his radical architectural theories, attended his dinners where, after sumptuous food, he would stricture them on prevailing upper-class taste. But beneath the surface a change was taking place, illustrated in the changing character of Sullivan's commissions. In the Auditorium, he found a first great outlet for his talent precisely because of the building's social usefulness. The theater was to be a music center for a music-loving population. It became in fact the musical center of the mid-continent, a pump from which flowed a whole stream of stirring cultural impulses. But there were no such implications in the Schlesinger and Meyer Building of 1899. For all its polished perfection, it remained a department store; and such a building of necessity sets sharp upper limits on social significance. In 1873, Louis Sullivan had found Chicago exhilarating; but as the century drew to a close, it was little short of explosive.

Ultimately Sullivan foundered upon this dichotomy, both as man and as architect. The Columbian Exposition of 1893 was the first intimation of his defeat. He had steadily developed his thesis of a democratic architecture, of genuinely *popular* esthetic standards which contradicted the realities of neither technical nor political progress. The architects, he had argued, must ally themselves with the people, learning from them as well as teaching them. "Are you," he had asked, "using such talents as you possess for or against the people?" He was convinced (and he was largely right) that whatever success he and the Chicago School had enjoyed was due to this philosophy. And he assumed that they would fight for it as determinedly as he. Here he was wrong. The ease with which Chicago architects dropped one set of esthetic standards to embrace another and much lower one was proof of the underlying political content of all concepts of "beauty." The first sign of betrayal came to Sullivan at the opening session of the Architectural Commission for the Exposition. Daniel Burnham was chairman of the group, which was heavily weighted with Easterners. While George B. Post (architect of the Pulitzer Building), Richard Hunt (architect for the Astors, Vanderbilts, and Goelets), and Charles McKim (architect for the Tiffanys) settled themselves around the table, Burnham began to speak. And Sullivan noticed in amazement that Burnham "was progressively and grossly apologizing to the Eastern men for the presence of their benighted brethren of the West."

Daniel Burnham, whose firm had previously been associated with the Chicago School, supported the Easterners when they proposed that the Fair be designed in the Roman style. Sullivan heatedly opposed this, arguing for the application of those esthetic standards for which the Chicago School was justly famous. But when the ballots were in and counted, the Classic had carried the day. This was clearly the approved symbolism of Wall Street, as every architect on the Fair Commission could easily have testified, and its appearance in Chicago in 1893 was more than an accident of changing public taste. Louis Sullivan, with only the design of the Transportation Building to worry about (Fig. 167), saw the interconnection of events:

> During this period there was well under way the [trend toward] formation of mergers, combinations, and trusts in the industrial world. The only architect in Chicago to catch the significance of the movement was Daniel Burnham, for in its tendency toward bigness, organization, delegation, and intense commercialism, he sensed the reciprocal workings of his own mind.[17]

Young Frank Lloyd Wright, asked by Daniel Burnham to join his firm just about this time, confirms Sullivan's estimate of Burnham's choice. Urging Wright to stay with the eclectics, Burnham told him:

> The Fair is going to have a great influence in our country. The American people have seen the "Classics" on a grand scale for the first time. . . . I can see all America constructed along the lines of the Fair, in noble, dignified, Classic style. *The great men of the day all feel that way about it — all of them.*[18]

The Chicago World's Fair was a shattering blow at the only consciously progressive effort of the century to resolve the growing contradiction between the ideology and the mechanics of American building. Historically, this was perhaps inevitable; but the loss was nonetheless disastrous. Referring to this period in post-Fair Chicago, Wright said in 1940: "They killed Sullivan and they nearly killed me!" And the defeat of this brilliant and erratic man (for though he lived until 1924, Sullivan's career was substantially ended by 1900) was, as Sigfried Giedion has so succinctly put it, due to the fact that "American architecture was [being] undermined by the most dangerous reaction since its origin."

What was the *source* of this reaction? Was it merely, as Professor Giedion would have us believe, the architect's desire to give "an artificial backbone to people who were weak in their emotional structure"? Many things might have been said of Chicago's beef and railroad barons, but scarcely that their "emotional structure" was "weak." So abrupt a change in esthetic standards was not to be

Fig. 167. "Golden Door," Transportation Building, Columbian Exposition, Chicago, Ill., 1893. Adler and Sullivan, architects. This single commission gave Sullivan his only opportunity to demonstrate esthetic alternatives to the official neoclassic idiom adopted by the Fair "establishment."

Fig. 168. Westinghouse Exhibit, Columbian Exposition. The patterns of dynamo and steel arches are echoed with uncanny (if unwitting) accuracy by the concentric arches of Sullivan's golden door (above).

explained in terms of esthetics, but of basic changes in American society. The end of the century saw the substantial completion of the modern structure of monopoly and its absorption of the Chicago capitalists. When they exchanged local partnerships for national trusts, they did more than acquire Wall Street's stocks and bonds; they also exchanged the last remnants of their provincial democracy for Wall Street's ideology.

THE WHITE CITY: "TRIUMPH OF ENSEMBLE"

The motives of the men who decided that the Columbian Exposition would be built in the Roman style may thus be explicable. But this does not of itself account for the immense popular success of the Fair or for its subsequent effect upon American taste. Most of the millions who visited it were, after all, ordinary middle-class, middle-western Americans with neither great wealth nor real power. Though some of them might be swept off their feet by the chauvinism of the Spanish-American War a few years later, it is hard to imagine that any of them had any close identification with imperialism, past or present. And certainly the knowledge of Latin literature and Roman art which had been an integral part of American culture in Jefferson's day had been long ago dissolved by the Victorian interregnum. Hence the success of the Fair must have been due to other factors than merely the particular architectural metaphor in which it was developed.

From contemporary accounts it is clear that the Fair left Americans dazzled by a totally new concept of urban order. It placed before them a set of images such as only those few who had visited Haussmann's Paris or Franz Joseph's Vienna might have known: polished squares and boulevards which were paved, landscaped, lighted, and *finished.* Most Americans, even those from the older Eastern cities, would have never lived on a street completed from one end to the other. Fifth Avenue or Michigan Avenue might boast a few blocks of tidy cosmopolitan splendor but they would quickly peter out into raw subdivisions or open farmland. And most Midwesterners would have grown up in cities which had quite literally grown up around them, subject always to the violently unequal

Fig. 169. Palace of the Mechanic Arts (above) and Administration Building (below), Columbian Exposition. Sketches like these convey, even better than photographs, the excitement caused by the landscaped splendor of the Fair. It was the first great display of outdoor electric lighting.

stresses of rapid urban expansion. The very urbane amenities which mark the modern street were becoming a reality only in the decade of the Fair. A wide use of outdoor illumination by electric light, for example, was actually first demonstrated at Jackson Park. It drew ecstatic reactions from people who had seldom seen a lighted street, much less a lighted fountain.

Most commentators agreed that the "White City" on the lake opened up new possibilities of order and urbanity for American cities. The novelist William Dean Howells thought its chief beauty derived from its symmetry. Henry Adams praised it for "imposing classic standards on plastic Chicago." John J. Ingalls said that, with the Fair, Chicago was no longer provincial: "She has established her claim to take first rank among the great capitals of the world."[19] So struck were most observers by its beauty and so worried by its transience, that speculation immediately began about the possibility of preserving it. "These enormous, splendid palaces of a day . . . announcing, as they do, the birth of a new art . . . must vanish tomorrow: but why should their durable counterparts not be reared?"[20]

Why not indeed? Daniel Burnham, with the magisterial complacency of the Director of Works of so successful a venture, had no misgivings:

> The influence of the Exposition on architecture will be to inspire a reversion toward the pure ideal of the ancients. We have been in an inventive period, and have had rather contempt for the classics. Men evolved new ideas and imagined they could start a new school without much reference to the past. But action and reaction are equal, and the exterior [sic!] and obvious result will be that men will strive to do classic architecture . . . designers will be obliged to abandon their incoherent originalities and study the ancient masters of building . . . The people have the vision before them here and words cannot efface it.[21]

The prospect struck Montgomery Schuyler as "not a promise so much as a threat." In an uncannily prescient essay, *Last Words About the World's Fair,*[22] he wondered

> what influence the display at Chicago is likely to have upon the development of American architecture, and how far that influence is likely to be good and how far bad. That it is likely to be in any degree bad is a proposition that be startling and seem ungracious, but there is no reason

why it should. Certainly to question the unmixed beneficence of its influence is not to pass the least criticism upon the architects, the brilliant success of whose labors for their own temporary and spectacular purpose has been admitted and admired by all the world.

Wise architects might understand that the architecture of the Fair was not for the common light of day. But many architects were not wise; and upon them "the buildings in Jackson Park are more likely to impose themselves as models for more or less direct imitation . . . such an imitation can hardly fail to be pernicious."

Schuyler then went on to point out that the success of the Fair was not due to the classic style that the individual buildings employed at all. On the contrary, "the success is first of all a success of unity, a triumph of *ensemble*. The whole is better than any of its parts and greater than all of its parts, and its effect is one and indivisible." And then he penetrates to the very heart of the problem. *"The landscape plan is the key to the pictorial success of the Fair as a whole."* It "generated" whatever detailed architectural success there was by supplying guide lines which "sensitive architects had no choice but to follow." This, and this alone, was the aspect of the Fair worthy of emulation. To mistake its architecture for "the actual or the possible or even the ideal architecture" of American life would be, Schuyler felt, disastrous. For the task of architects was not

to produce illusions or imitations, but realities, organisms like those of nature. It is in the "naked and open daylight" that our architects must work, and they can only be diverted from their task of production by reproduction. It is not theirs to realize the dreams of painters.

But it was Burnham's advice, not Sullivan's or Schuyler's, which was to carry the day. It may have contributed somewhat to raising popular comprehension of municipal esthetics (the rehabilitation of L'Enfant's plan for Washington was a direct outgrowth of the Fair). But against architecture itself it constituted a counterrevolution from which it would take half a century to recover. It drove into exile or eclipse those men whom it could not seduce. It served to discredit or drive underground the broad stream of esthetic rationality and social accountability that marked the theories of Greenough, Emerson, Bellamy, and Sullivan. It entombed American architecture in the coldest, most inflexible dogma of its history.

7. 1893–1933

ECLIPSE

For both Louis Sullivan and Frank Lloyd Wright the Columbian Exposition was to prove climactic, a turning point in their lives. Both had refused to be a party to the Classic hoax and this courageous decision had implications which neither one could have fully foreseen. As the new century opened, Sullivan faced eclipse, whereas Wright stood upon the threshold of the most remarkable career in American architecture. In tracing Sullivan's descent into obscurity, it is obvious that both external and internal forces were involved. That there was a change in his cultural milieu is indisputable. But in 1899, at the age of forty-three, he must have seemed to be standing at a pinnacle of power and prestige. The Bayard Building in New York was just finished, the Gage and the Schlesinger and Meyer Buildings were under way. By all conventional standards, he would have been reckoned a successful practitioner, enjoying the confidence of State Street bankers and merchant princes.

Yet he was already beginning to formulate *The Kindergarten Chats*. The programmatic character of this attack upon eclecticism and the pharisaical society of which it was a visible manifestation shows that he had given long and careful consideration to the defeat he had suffered at Burnham's hands eight years before. The tone of these essays is often one of mordant good humor. But the sardonic ridicule of those bankers and businessmen who were commissioning Roman banks and skyscrapers is merciless. If, as seems probable, he was speaking publicly in the same vein, it need not surprise us that those gentry began to shy away from him.

Perhaps he was under the impression that he could still attack such sensitive areas with professional impunity. His entire career, after all, had been structured upon artistic nonconformism and intellectual dissent. If this was the case, then he did not yet understand how literally the decision at the Fair was to be interpreted as the end of an epoch of progressive experimentation. The Schlesinger store was the last great commission of his life and the last but one of his Chicago projects. He had been the architect of some 106 buildings in the preceding twenty years. He would do only eighteen more in the twenty-five years which remained to him.

A few of his clients after 1900 must have been congenial. He did the finest house of his whole career for the Henry Babsons in Riverside in 1907 and, in the following year, the lovely bank building at Owatonna. In both cases, he was working for people who sought in their architecture a direct and rational expression of a cultivated but modest way of life. In the case of L. L. Bennett, the Owatonna banker, the search was quite explicit. He had rejected "the classical style of architecture so much used for bank building," he wrote in 1909,[1] "as being not necessarily expressive of a bank . . . and defective when it came to practical use." He had sought — and found in Sullivan — "an architect whose aim it was to express the thought or use underlying a building, without fear or precedent."

Fig. 170. Side view, National Farmers' Bank, Owatonna, Minn., 1907–08. Louis Sullivan, architect.

Fig. 171. Entrance seen from Banking Room, National Farmers' Bank, as restored in 1958 by Harwell H. Harris. The original plan (below) included rental shops along the side street and second story offices which bridged the alley.

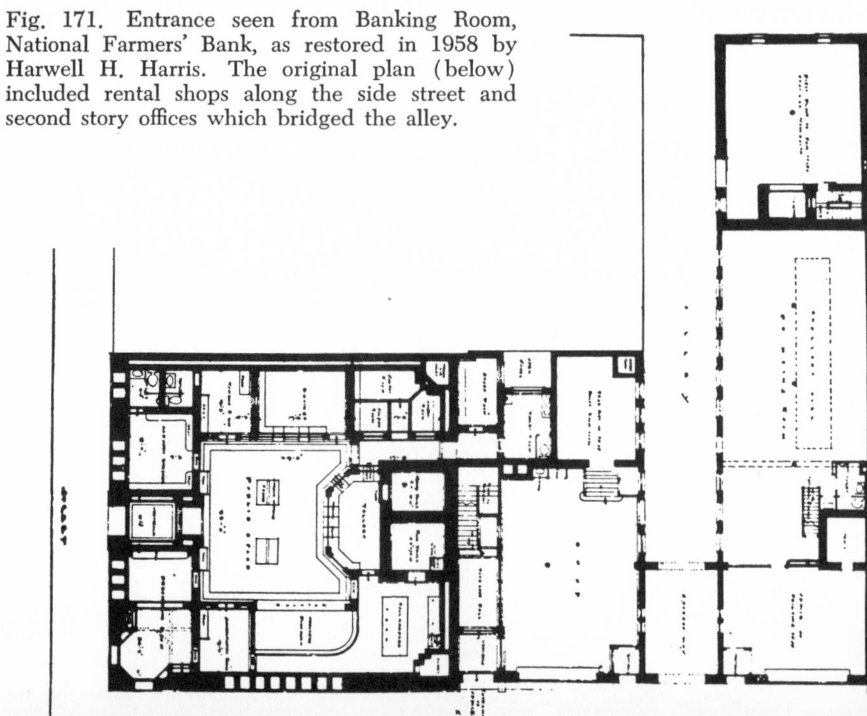

Fig. 172. Entrance to Vaults from Main Banking Room as designed by Sullivan.

Fig. 173. Same wall as redesigned by architect Harris. All the brilliantly polychromed plaster ornament was carefully restored, as were Sullivan-designed chandeliers and wall paintings in lunettes.

But such clients as these were few and far between: not enough to deflect the defeat which engulfed him. In 1906, his wife divorced him under circumstances as obscure as those which had led to their marriage in 1899. Then followed the loss of his Gulf Coast home in 1908, the departure of draftsmen one by one, the auction of his art and furniture in 1909, withdrawal from the Chicago Club in 1910, his retreat to a wretched hotel in 1911.[2] This drop into threadbare poverty must, of course, have been compounded by his own internal conflicts. These are diagrammed with incandescent anger in his manuscript *Democracy: A Man-search.*[3] Written in the summer of 1908, these essays reveal him as anything but reconciled to his society:

> . . . for to see our feudal and barbaric civilization naked, is to see a horror, a ghost; a Dance of Death.

He has lost all confidence in the rich, the learned, the mighty:

> . . . the eminent will not, cannot, do not know how, and persist in not knowing how [to change the course of events]. There is neither health nor help in the eminent.

His only faith is in Democracy:

> The people will awaken whenever the veil of feudal illusion, rotting with the over-ripeness of time, drops down from their eyes [exposing] the naked ferocity which they now sanction, and in which they are enmeshed.

But illness and old age had dimmed this apocalyptic view of American society when Sullivan came to write his *Autobiography* in the last year of his life. Ironically describing in the third person that earlier Sullivan, he says that

> of politics he knew nothing and suspected nothing, all seemed fair on the surface. . . . He had heard of the State and read something about the State but had not a glimmering of the meaning of the State. He had dutifully read some books on political economy . . . and had accepted their statements as fact. He had also heard vaguely something about finance and what a mystery it was. In other words, Louis was absurdly, grotesquely credulous.[4]

It may well be that no single man could have breasted that Classic tide. The architect of Louis Sullivan's day was much more the creature and prisoner of the wealthy than he is today. Even Frank Lloyd Wright, for all his splendid display of nonconformism, did not really escape. He merely found patrons who were themselves rich nonconformists. But Sullivan could scarcely have succeeded even if objective conditions had been different. His concept of democracy was a cloudy thing. It is significant that though he envisioned an abstract humanity moving in an abstract architectural felicity, improving and improved by a democratic esthetic, he failed utterly to concretize that vision.

He believed that "in our study of social science altogether too much importance has been attached to heredity and too little to environment." Yet he was never able to align himself with those people or movements which sought to change the environment. For all his interest in the application of science and technology to human welfare, he submerged himself in cloudy metaphysics. He was convinced that "the so-called average mind has vastly greater powers, immeasurably greater possibilities of development than is generally believed." Yet in his personal life he showed no interest in average people. He grew personally autocratic and egotistical, more and more isolated from the family of men. He had no allies when he met the aristocratic invaders from the East on the board of the Columbian Exposition; and the defeat he met at their hands did not drive him into closer contact with the people. On the contrary, it had the tragic reverse of driving him into isolation, abstraction, mysticism. Years before his death in April, 1924, a great figure had been lost to American building.

THE GREAT DISCIPLE

If Frank Lloyd Wright succeeded where Sullivan could not, one might almost put it down to his extraordinary tenacity. His accomplishment was, as he himself put it in later years, in simply staying alive. Certainly, in a long and turbulent career, there were many times when it must have looked as though he had gone down for good. Thus his career might well have been assumed to be finished when he left his wife and children to flee to Europe with Mrs. Cheney in 1909. Or again in 1914 when, on a day of Walpurgian horror, a mad servant had murdered her and her children and burned the newly finished Taliesen down around their corpses. And he did in fact disappear from the American scene the next year, when he was invited to Tokyo to design the Imperial Hotel. That project kept him abroad almost continuously for five years; and when he did return, it was not to his old habitat in the Midwest but rather to Southern California. There, in a new burst of creative response to a new landscape and climate, he did a series of remarkable houses. Today these wonderful projects from his so-called "Mayan period" are ranked alongside his greatest work. But they attracted little attention at the time. The Establishment must have reckoned him safely buried, along with the troublesome ghost of Sullivan. Even so brilliant an observer as the young Lewis Mumford paid them scarcely a passing reference. As he wrote much later:

> The story of American architecture after 1850 was a pathless waste. By 1924, the work of the Chicago School, historically, had dropped out of sight completely. This means that for the historian the most creative period in American architecture, that between 1880 and 1900, did not yet exist.[5]

But Wright had prepared an enormous surprise for those who prematurely declared him dead. Not only was he to re-emerge upon the stage of American architecture as its greatest single figure, enjoying decades of prestige and *éclat*. He had also set in motion a great movement which, already in the twenties, was remaking the face of European architecture; and which, crossing the Atlantic in 1932, would overturn and completely bury the creaking eclectic apparatus of the American Establishment. Of course, no single individual ever created a "movement" single-handedly. But historians on both sides of the Atlantic are today agreed that Wright's work before World War I played an absolutely pivotal role in the development of modern architecture. The first European publication of his work in 1910 had been of incalculable influence upon Gropius, Mies, and the men of their generation. But even before that, his work had been known and praised in Western Europe. Kuno Francke, the German critic, had traveled to Chicago to see him in 1908. And H. P. Berlage, the great Dutch contemporary of Sullivan, had gone to see the Larkin Building and come away "with the conviction of having seen a genuinely modern work, and with respect for the master able to create things which had no equal in Europe."[6] In fact, American architects in the twenties seem to have been the only ones on earth who were unable to see the significance of what Wright had done.

In retrospect, it is apparent that Wright completed the task begun by Richardson and carried forward by Sullivan in a manner all too rare in cultural history. They form a continuum in time and space, each the indispensable precondition for the appearance of the next. However, heir though he might have been, Wright did not merely sit back to enjoy his heritage. He used what they gave him, but he went far beyond that. With uncanny consistency he hewed to his own line. He was never isolated from the world around him, yet never submerged by it. In magazines, books, luncheons, classrooms, and most of all, in his designs, he stated and restated his convictions. Was there a need for the civilized rich to break away from the formalized barbarism of Hunt's great house for the Vanderbilts? Wright's house for the Coonleys was the prototype of the break. Could no American architect conceive an urbane outdoor eating place? In the Midway Gardens on Chicago's lake front Wright showed them how to do it — a genuinely gay and lovely place, with lights, flowers, and music built in. Could mighty American business provide no more healthy or attractive an environment for its clerks than the dark warrens of its office buildings? Look to the Larkin Building in Buffalo. Could a religious congregation whose creed was rational and liturgy simple find nothing more suitable than the current ineptitudes? Wright's Unity Temple in Chicago gave them a functional assembly room, free of archaic symbolism, full of light and air.

The examples of his perception and foresight, the range of his interests, are unmatched since Jefferson. If he was less profound as a thinker, he was a far more brilliant architect. Of *plan* he was from the start the master; in *structure* no less brilliant, though more erratic; in *mechanics* and *equipment* always forward-looking and adventurous, ready to give any new heating system or building material

Fig. 174. Fredrick C. Robie house, Chicago, Ill., 1909. Frank Lloyd Wright, architect. Here the pinwheel plan of the Prairie house has been brilliantly modified to fit a long narrow lot and the floor levels manipulated to win both privacy and view on the suburban street.

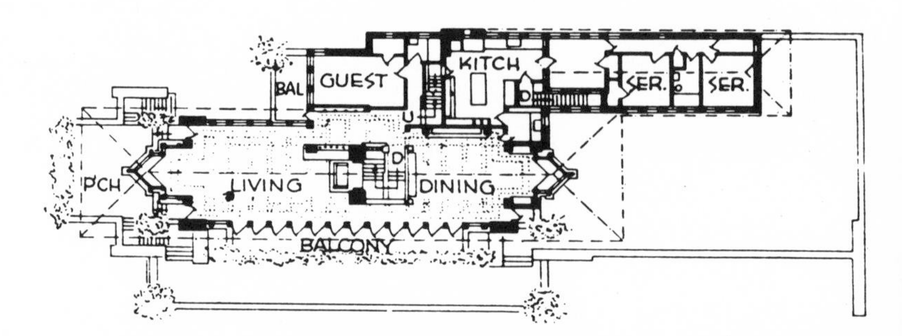

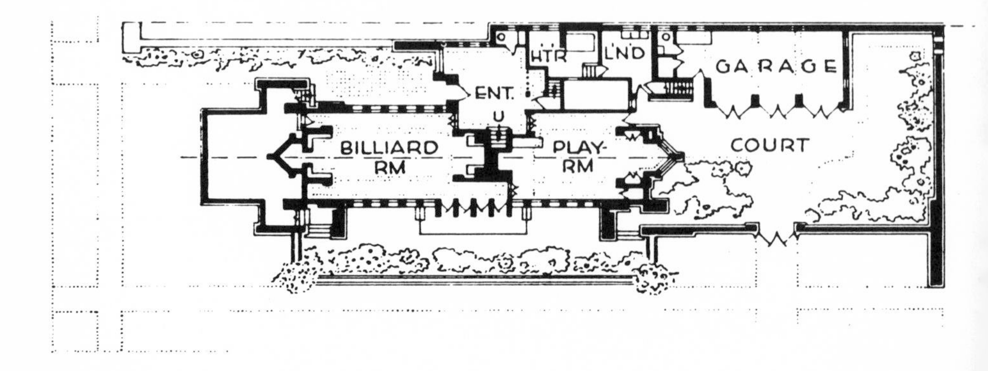

Fig. 175. Living room, Coonley house, Riverside, Ill., 1908. Frank Lloyd Wright, architect. His theory of organic ornament is demonstrated in the ceiling of this great room, whose wooden patterns are an abstraction of its structural framing.

a whirl. But it is in the field of sheer *esthetics* that he is unique. In this specific area he was for half a century an incendiary, putting outworn prejudice and accumulated historical bric-a-brac to the torch.

Critics have for some years now been busily cataloguing his designs, indexing his "sources," defining his "periods." As for all great artists, this is essential labor — and difficult, for Wright was a man of enormous and complex culture. But the distinct stages of his development should not obscure its continuity. For the quality which ties all the vast quantity of Wright's work together, separates it from his imitators, and raises it above his detractors, is his highly personal (one might almost say, private) esthetic standard. Although descended from Sullivan's, it is clearly not the same. Sullivan's motivation was in expressing and elevating the standards of taste of the undifferentiated masses — Wright's in deepening and enriching the perception of the individual. Sullivan wrote of the mass perversion of public taste; Wright has dwelt at length on the effect of boyhood summers on his grandfather's farm upon his concepts of beauty. One sought to be a democrat, the other was a humanist.

And this preoccupation was from the start apparent in Wright's houses (Fig. 174). Gleefully, he swept out the trash. He stripped the cornice of its brackets and the roof of its dormers; he lopped off the phony cupolas and the gingerbread; he standardized the windows and swept them almost brutally into large horizontal banks. Tortured surfacing materials gave way to sheer texture. Everywhere, a masculine and electric purpose was apparent. Indoors the same process was afoot. Dictating, and not always wisely, the design of even the fabrics, the architect swept out muddy frivolities and replaced them with body-warm colors of coral brick, sand plaster, natural oak, and raw linen. Furniture became honest, perhaps too harsh and militant, but antiseptic by contrast. Light fixtures were first rationalized into geometric patterns, then built right into the fabric of the house. Whole walls disappeared and rooms ran together, until a respectable person could scarcely say where indoors stopped and outdoors began. A fresh wind blew through American houses, and sixty years later the dust has not yet died down.

Fig. 176. Unity Temple, Oak Park, Ill., 1904. Frank Lloyd Wright, architect. To overcome the disadvantages of a busy, downtown corner, Wright developed a novel plan (below). A raised central terrace and foyer provide sheltered entry to auditorium and Sunday School. The inward-turning, top-lighted auditorium is raised above street level, insulated against noise and distractions (section, bottom).

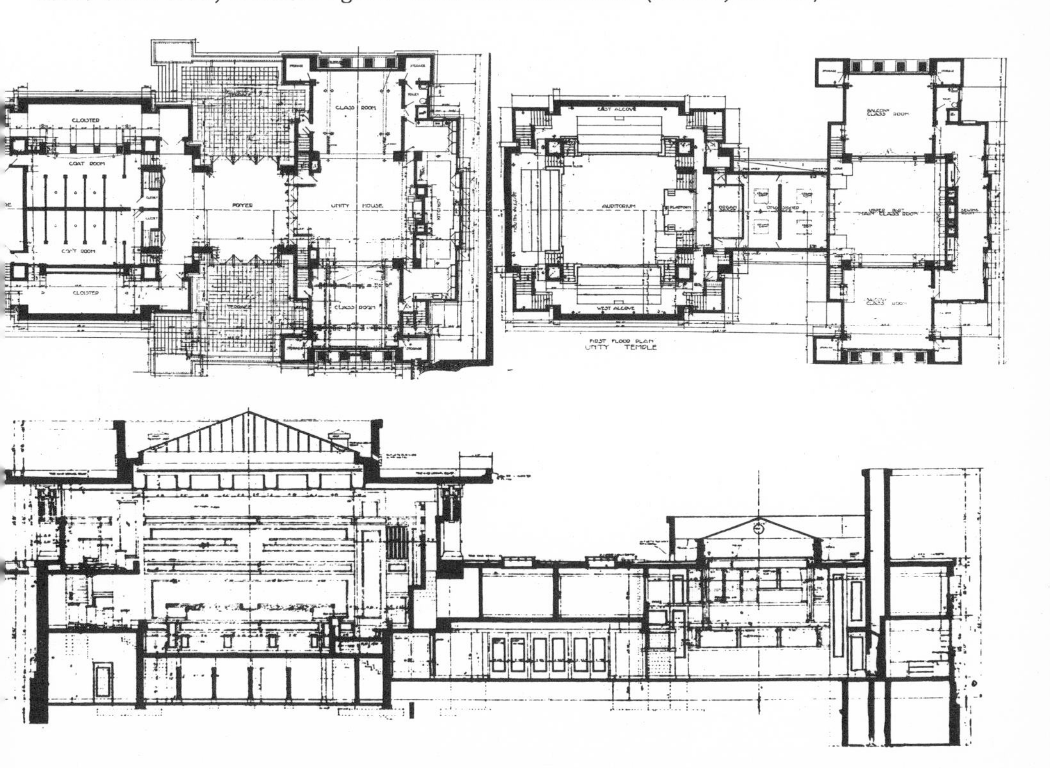

Wright's purpose, his whole palette and skill, was to deepen and enrich the esthetic experience of the individual. It is remarkable that so much has been said about the angularity and harshness of his houses; so little about their sensuality, their direct and immediate appeal to the senses (Figs. 175, 186).

Wright's professional career was a paradox. It was a constant struggle with the Blimps and the Plushbottoms: he never won, but he never lost. Just when the latest Wrightian eccentricity had been safely interred at some professional dinner, Wright reappeared with a new commission or another project with which to startle his colleagues and intrigue the literate public. Contrariwise, by some tragic perversity, his enemies were able repeatedly to use his personal life as a weapon against him. His separation from his first wife, in a day when divorces were *outré;* the murder of his second wife; his personal flamboyance; his bankruptcy — none of these strengthened his position with boards of directors, school committees, and municipal councils upon whose goodwill the architect must largely depend.

Yet he succeeded where Sullivan had failed. Above and beyond his inexhaustible vitality and immense talent, this was also attributable to the fact that he never overextended himself, as Sullivan had done. Though he wrote and lectured a great deal, it was always to explain his architecture; never, as with Sullivan, to remake the whole world. In this sense, he was far more realistic in his commitments than his *lieber Meister.* He remained at base an incorruptible individualist. Though he reacted violently to society's concrete pressures upon him, he never showed much interest in society in the abstract. Only once, in his *Project for Broad Acres Town* (1935), does he engage in social theorizing or Utopian designs. Physically, the project might have produced a handsome suburban village. But socially it is a typical product of the depression years: a *petit bourgeois* Utopia, an odd mixture of subsistence farming, cottage industries, power tools and white collar commutation to some nameless metropolis.

Wright was never neutral, however. Toward such issues as war and peace, divorce and marriage, land speculation, and intellectual conformism, his reactions were always prompt and uncompromising.

Fig. 177. Larkin Building, Buffalo, N. Y., 1904. Frank Lloyd Wright, architect.

Fig. 178. Midway Gardens, Chicago, Ill., 1913. Frank Lloyd Wright, architect. In these two vanished buildings, Wright displayed a maturity so assured and complete that they seem exactly contemporary with buildings he was to do fifty years later.

But they were always personal, "unorganized": sometimes, they appeared simply idiosyncratic or perverse. Yet throughout a long life, Wright preserved an uncanny balance. His personal courage often carried him into dangerous waters where, like a convoy dodging torpedoes, a zigzag course was essential to survival. With a tenacity rare among mortals, he never lost sight of his objectives and these made him into our country's greatest architect.

Frank Lloyd Wright's influence upon American building design has been immense, incalculable; for that reason it is also hard to assess. In his buildings, it has been generally progressive. The whole average level of our architecture, especially our domestic architecture, is incomparably higher than it would have been had Wright not lived.

THE WASTELAND

At first glance, the period from 1900 to 1933 appears to be an esthetic wasteland: and a closer scrutiny of the individual buildings of the period does little to correct that first impression. Those structures which were hailed as the landmarks of the time — the Woolworth Tower, the San Francisco Exposition, the Mizner storybook houses in Florida, the *Chicago Tribune's* Gothic extravaganza, the Triangle in Washington: these buildings are so colorless and without significance, so magnificently sterile, as to leave one nonplussed. There may have been a certain bleak majesty in the scale of the skylines which were rising all over America. And there was a spurious romance to the suburbs of the wealthy in their shadow (Fig. 204). People with "good" taste were trapped like flies in honey in the mellow compositions of Mellor, Meigs and Howe, in Philadelphia; they worshiped in such steel-and-plaster vaults as the Riverside Church afforded; they celebrated the spread of culture in the towering pile of the Chicago Civic Opera.

But beneath all this empty and eclectic froth there was, of course, an enormous quantitative expansion in the building field. It had taken whole decades for the skyscraper to develop in the hands of Jenney and Sullivan. But now it sprang up in full bloom across the land, changed hardly a jot or a tittle from the Wainwright Building in St. Louis. Civic centers in the cold gray stamp of dead Romans appeared at hardly more than the press of a button. Apartments, factories, schools, and churches — never had a people built so many in so short a time. However badly planned and atrociously ornamented, they were for the most part remarkably well built, well heated, and well plumbed.

This was the great period of rationalization in American building technology; its importance must not be underestimated. It is one thing to build a single thirty-story skyscraper in a big city. It is quite another to be able to reproduce them at will across the land. This required the raising of the *entire average level* of the building field. The real impact upon American life of such developments as the elevator came not with Otis's demonstration at the New York Fair in 1853, but rather with the wide appearance of multi-storied buildings at the turn of the century. As early as 1836, Tredgold's treatise established the basis for steam heating: yet almost two-thirds of a century passed before small-town heating-and-plumbing men were able to put them into widespread use. None of the brilliant experimental work of Volta, Davy, and Franklin could mean much to the man in the street until Edison, in the seventies, had perfected the incandescent bulb and the central generating plant.

So it was in every section of the building field. A century of slow handicraft accretion gave way suddenly, like a log jam, to the full flood of industrialized mass production. For three decades or more — from around 1900 to 1930 — the building field was characterized by quantitative advance along well-established lines of endeavor. It was perhaps a necessary gestation period, during which the skin of hopelessly archaic esthetic formulas were stretched to the bursting point. No building was too tall for a Classic colonnade, no gymnasium too bulky for "Collegiate Gothic," no house too small for false Elizabethan timberwork.

Recent historical research[7] has rediscovered several minor progressive currents concealed beneath the surface of this frozen Martian landscape. A post-Wrightian Chicago School was one such, a loosely associated group of Southern Californians was another. However, neither the esthetic significance nor the extent of their influence should be exaggerated. The fact that they survived at all is a measure of their vigor; but the fact that they had been all but suffocated by the end of World War I is an index of how total was the systematization of American taste under the rule of the Establishment.

Of the Chicago group, the most successful was George G. Elmslie. A talented protégé of Sullivan's, he had loyally stayed with him until 1909, when he left to form a partnership with William G. Purcell. In the following decade, they designed a wide range of buildings — homes, churches, banks and stores — which reflected quite accurately the rational and progressive milieu of their middle-class clientele. Their work was considered important enough to merit occupying two entire issues of *Western Architect* in 1915.[8] But even a cursory examination of this work shows it as derived, in about equal parts, from that of Sullivan and Wright (Figs. 179, 180). Sullivan himself, of course, lived on until April 1924: but in such deepening obscurity that a foreign admirer like the young Viennese architect Rudolph Schindler had some difficulty locating him.[9]

A number of Wright's associates carried on in his general idiom in the decade after 1910: Marion Mahoney, the first woman licensed to practice architecture in Illinois; her husband William Burley Griffin, who won the competition for the design of the new Australian capital of Canberra; William Drummond; George Maher. Their work was incontestably superior to the Beaux Arts mediocrity around it. But it lacked the power and creative range of the masters, without whose presence it withered and died. Thus it was that the whole Chicago School was entombed by regnant opinion — so completely so that as we have seen, the diligent and sharp-eyed Mumford all but passed it over in 1924.

SOME WEST COAST VARIATIONS

The rapid development of Southern California as a vacation and retirement area was due to a special kind of immigration from the East. The result was a demography unique in American history: communities of well-to-do and educated people, uprooted from normal routines, detached from the familiar European orientation and brought up against the unfamiliar Orient and embedded in a

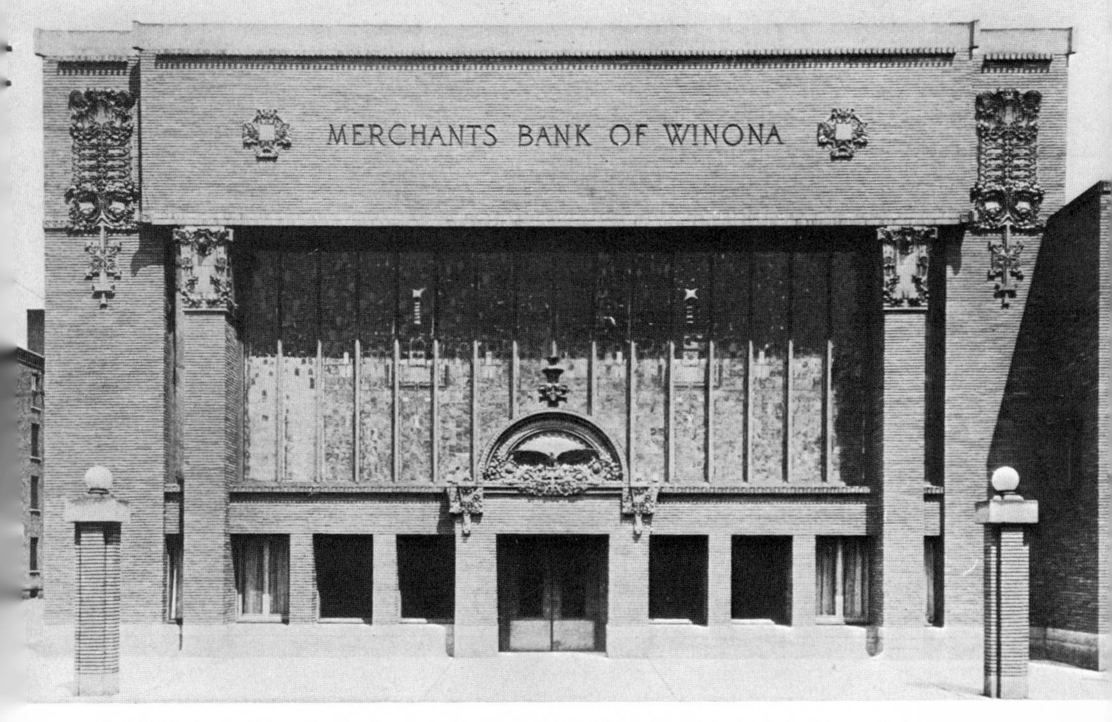

Fig. 179. Merchants Bank, Winona, Minn., 1913 (?). Purcell and Elmslie, architects.

Fig. 180. The Edison Shop, San Francisco, Calif., 1914. Purcell and Elmslie, architects.

genial climate and exotic landscape. The result was a milieu friendly to innovation of all sorts. Given the wide variety in background of the immigrants, it was inevitable that this innovation be heterogeneous intellectually and artistically. In the wealthier settlements like Santa Barbara and Pasadena a certain sobriety ruled. In nearby Los Angeles, any lunatic extreme was possible. But the central phenomenon of the whole region was the extraordinary plasticity of its taste.

It was in such a context that Charles and Henry Greene were able to build a series of significant houses in Pasadena between 1900 and 1915. Though these bear a superficial resemblance to Wright's houses of the same period, there seems to have been no connection whatever between them. Their similarity springs directly from the fact that both were strongly influenced by the Japanese. If today the Greenes' houses seem more dated than Wright's, this is doubtless due to the fact that they had not his majestic powers of transmutation: their use of Japanese precedent is tasteful but literal and direct. These houses, nevertheless, represent an important architectural invention, incorporating all the distinctive features of the West Coast house today (Figs. 181–182).

The Greenes' clients were usually wealthy Midwesterners of liberal Protestant or Quaker background. They belonged to that segment of opinion which supported national parks, woman's suffrage, progressive education, factory reform. They were involved in new theories of love and marriage, of birth control and child care, of diet and hygiene. At the practical level, these interests are more or less directly expressed in the twin beds, sleeping porches, children's rooms and well-planned kitchens of the Greene houses. And at the symbolic level, these values are clearly expressed in the hospitable openness, comfortable scale and warm friendly color schemes. The Greenes had control of the interior design and furnishings of many of these houses. This perhaps explains the absence of antiques and heirlooms, of family portraits and coats of arms — in short, of all the usual upper-class iconography of caste and status. Though they were comfortable, even luxurious, in their accommodations there is a deliberate lack of pretension in their design. If these houses resemble

Fig. 181. Stair hall, D. B. Gamble house, Pasadena, Calif., 1909. Greene and Greene, architects. Using West Coast woods in the same way as the Japanese, the Greenes raised carpentry to the level of cabinet making.

Fig. 182. Entrance front, Gamble house. Similarity with Wright's prairie houses seems to stem from common inspiration in Japanese art: there is no indication of any contact between the two offices.

Wright's, the reason is not far to seek: the Greenes were designing for the same sort of people as those for whom Wright was building his "prairie houses." Allowing for differences in climate, vegetation, and topography, they represent the same sort of adroit and sympathetic response to the clients' requirements.

Symbolically, the Greenes finished the last of these houses in 1915, the very year in which the eclectic tide reached California in the form of two big expositions — the California-Pacific in San Diego and the Panama-Pacific in San Francisco. Both were victories for the Establishment, though Western vitality managed to breathe a little compositional sparkle even into the Beaux Arts project in San Francisco. In San Diego the style resurrected by Bertram Goodhue, the elegant prestidigitator from New York, was the Spanish Colonial. The arguments for its resurrection were admittedly more plausible than usual. On the one hand, a rustic version of it had been the authentic vernacular of the region until the American seizure from Mexico in 1848. Many buildings from this period still stood in Los Angeles in 1900. On the other hand, like most folk architecture, such buildings were well adapted to environmental conditions: the characteristic patio plan, wide roofs, and heavy mud-walled construction were ideal responses to a hot semi-arid landscape, such as this area was before irrigation converted it into something quite different. Thus, the Spanish Colonial revival begun at San Diego did perhaps less violence to Montgomery Schuyler's "light of common day" than most exposition architecture. Be that as it may, however, it was quite as hostile to independent work like that of the gentle Greene brothers as was the Beaux Arts classic in San Francisco; both served to "organize" West Coast taste into rigidly defined channels of "good" and "bad" taste.

Frank Lloyd Wright had already finished the Barnsdall house when he came back from Japan for the last time in 1922. Then he proceeded to take command of the Southern California landscape with the same confidence that he had shown in both Illinois and the Orient. Nothing could have been more different from these previous

Fig. 183. California Building, California-Pacific Exposition, San Diego, 1915. Bertram Goodhue, architect. Here the new eclecticism took the form of the indigenous "Spanish Colonial."

Fig. 184. Rotunda and Peristyle, Palace of Fine Arts, Panama-Pacific Exposition, San Francisco, Calif., 1915. Bernard Maybeck, architect. Beaux Arts idiom was employed with virtuosity and elegantly executed.

theaters of operation than the spectacular arid land forms and brilliant skies of Los Angeles. As usual, Wright approached this new field of action without any impedimenta of preconceived solutions. Instead, he analyzed the tradition, terrain, and technical resources of the region. In his cool masonry houses of this period — with their tall ceilings, shaded courts and terraces, and carefully controlled fenestration — Wright is actually much closer to local tradition than Goodhue with his archeologically accurate detail. Neither the mud masonry of the Spanish hacienda nor the sculpture-encrusted rubble of the Mayans was available to him. Earthquakes had rendered both modes of construction illegal for public structures and unwise even in private ones: and both were too costly, anyway. So he turned to his most plentiful material, concrete, and fabricated it into its cheapest form, cast block. Then, using a warp and woof of steel reinforcing rods, he wove it into his famous tapestry block walls. The way in which beauty, comfort, and safety were simultaneously won is dramatized in the beautiful Millard and Ennis houses (Figs. 185, 186).

But Wright shifted his center of operations back to Taliesen as the depression deepened, leaving the cause of modern architecture in the hands of two young Austrian architects, Rudolph Schindler and Richard Neutra. Appropriately enough, both of them had worked with Wright, Schindler having been in charge of Wright's Los Angeles office before he returned from Tokyo. However, it is apparent from even their earliest work — Schindler's beach house for Dr. Lovell at La Jolla (1926) and Neutra's great residence for the same man (1929) — that these men derived stylistically from European, not American, circles. Here was none of the faltering empiricism of Irving Gill or of the free-wheeling San Franciscan eclectics, Bernard Maybeck and Willis Polk. Both of these men employ a mature and complete visual syntax, far more sophisticated than anything else in America — always excepting Wright. Actually, both of these men belonged much more to the period that followed 1933 than that which closed with it. And their continued activity in Los Angeles during the following decades (Schindler died in 1953) played a greater role in the modern movement there than is commonly accredited them.

Fig. 185. Mrs. George M. Millard house, Pasadena, Calif., 1923. Frank Lloyd Wright, architect.

Fig. 186. Charles Ennis house, Los Angeles, Calif., 1924. Frank Lloyd Wright, architect.

Fig. 187. Philip Lovell house, Los Angeles, Calif., 1929. Richard Neutra, architect. This house, recently reconditioned by new owners, shows the astonishing stylistic durability of Neutra's early work.

Fig. 188. C. H. Wolfe house, Catalina Island, Calif., 1928. R. M. Schindler, architect. The polished geometry of the street facade establishes both Schindler's grace as a designer and his Viennese origins.

NEW AVENUES OF ADVANCE

However sterile the architecture of the period, esthetically and technically, there was a whole series of developments in allied fields of city planning, parks, and housing which were pregnant with possibilities. When Frederick Law Olmsted completed New York City's Central Park in the sixties, he ushered in a whole movement. The project created a genuine sensation. There was nothing in the country which could approach it in scale, imagination, or competence (Fig. 147). Other cities began to imitate it, employing Mr. Olmsted as designer. His work was the substructure to the subsequent City Beautiful movement, which had its specific architectural origins in Olmsted's own design for the overall layout of the Columbian Exposition. Largely an upper-class expression of civic consciousness, it gained further impetus from the rediscovery of L'Enfant's plan for Washington and the publication in 1901 of the report of the McMillan Commission for the reconstruction of the capital city. These drawings, envisioning long, empty avenues of carefully scaled elms backed by Classic façades, introduced to city councils all over the nation totally new concepts of order and civic majesty.

There was a lot of Baron Haussmann and precious little democracy in these vast geometries of befountained plazas and intersecting boulevards. Light, air, and foliage were concentrated around hypothetical water gates and imaginary civic centers, never around housing or schools. Few of these schemes had any relation to the basic needs of the community. They all wore the obvious air of being schemes to deceive visiting notables, like the elaborate screens allegedly put up along the journeys of Catherine the Great, to hide the actual wretchedness and confusion of the land behind.

Yet they played a vastly important role in introducing the concept of *planned* reconstruction into the popular mind. However pompous and autocratic the solutions, they were at least admissions that real problems did exist, and that local governments did have the power to exercise some sort of control over the urban environment. Moreover, as the movement matured, it involved many well-intentioned souls who were forced by the struggle itself to realize that the problem was much more than one of simple face-lifting. The movement was also to gain important accretions from the ranks of the social and welfare workers, who were increasingly aware of the relation between the physical environment and crime, delinquency, and ill-health. Their insistence upon such factors had a salutary effect upon the movement as a whole, and playgrounds, schools, and clinics began to appear alongside ornamental drinking fountains and statues of Spanish War dead.

But such movements were actually peripheral to the central fact — the crisis of the city itself. Unparalleled expansion during the last half of the nineteenth century had doubled and quadrupled the size of the older cities and produced dozens of new ones. "Rapid transit" — trolley, subway, elevated and steam commuter trains — had made this expansion possible. Horse-drawn vehicles had merely supplemented the transportation network — filled the interstices, as it were. Essentially radial, this pattern was relatively stable because of the very nature of rail and roadbed. It was almost indefinitely expansible but in relatively fixed directions.

It remained for the automobile, in three short decades, to reduce this already inadequate pattern to a shambles of confusion. The motorcar proved a disastrously flexible means for moving people, ideas, and things around the surface of the earth. It brought fantastic increases in vehicular traffic into the streets — traffic which had been hitherto tightly channelized into steel rails. In this way it profoundly altered the character and function of the street; on a city-wide scale, it made obsolete the entire street system.

A contradiction appeared. While the city as a whole continued to expand by centripetal attraction, the population within the city was flung, as though by centrifugal force, to a constantly expanding perimeter. Like cream in a separator, the wealthy and the middle class were flung to suburbs ever more distant from the center; and the rest of the population, in strict accordance with American housing tradition, moved into the backwash of their discarded homes. The automobile was a basic factor in all of this and left in its wake an urban turmoil of social, political, and economic problems. Land values, traditionally graded upward toward transportation and

Fig. 189. Two views, proposed reconstruction of The Mall, Washington, D. C., 1902. One direct result of the Columbian Exposition of 1893 was the rehabilitation of L'Enfant's plan for the capital city by a Senate commission headed by Sen. James McMillan. Daniel Burnham, Frederick L. Olmsted, Jr., Charles F. McKim and the sculptor Augustus St. Gaudens, prepared these designs, which were to govern the style of official architecture for the subsequent forty years.

downward from the center out, had to be adjusted to a means of transportation which was *everywhere*. Values rose to preposterous levels at the center and along the expanding rim, while dooming the intervening "blighted" areas to decay. Such a condition naturally played ducks and drakes with any rational pattern of land use. A given block in the path of urban expansion could easily be open farmland, stylish suburb, respectable boardinghouse, and unregenerate slum — all in the span of a decade or two. It imposed increasing burdens on the whole fabric of municipal government — sanitation, police and fire protection, utilities, schools. The fabric would scarcely be complete before the tax income would begin to drop below operating costs, and the municipality would be faced with the task of supplying the same services to a whole new ring of suburbs.

Such a state of affairs penalized everyone in the community, even if in varying degrees. Naturally the waste was most readily expressed in dollars and cents; and the landowners were among the first to agitate for relief. The city councils, always responsive to the real estate interests, began to adopt the technique of zoning as a means of controlling urban growth. There were two aspects to zoning. One was directed at maintaining some minimum standard of light and air in congested areas and among tall buildings. The other was directed at control of land use, so that an area's property values, based upon one type of use, would not be destroyed by invasion of another. Zoning had its limitations, chief of which was its essentially static intention to stabilize property values; but as it came to be integrated with the city plan it revealed other possibilities. It was directly responsible for the spectacular "setback" design of modern skyscrapers. More important in the long run was its tendency to establish certain minimal standards in terms of light, air, space. The effect of zoning upon housing, schools, and office buildings was in general progressive.

But external relationships *between* buildings are intimately related to spatial relationships *inside* the individual building. What was gained by pushing buildings back from street and property lines if there were to be rooms inside the building which had no access to outside light and air? What good were public sanitation systems if individual landlords were to be allowed to keep privy and pump? How could the fire departments, even with their newly motorized equipment, ever overtake the threat of spawning firetraps?

Another whole area of social control was involved here, an area as controversial as socialized medicine today. For it clearly implied subordination of the individual's right (to do as he pleased with what he owned) to the well-being of the majority. Disastrous fires and epidemics served constantly to emphasize the necessity for such control. Gradually building laws and ordinances were passed, building inspection departments established. These were both bitterly contested and openly flouted — even though, for the most part, these controls acknowledged the *status quo* in existing structures and applied only to new construction and remodeling. Buildings erected prior to the laws would have to be obsolete, literally unfit for human habitation, before the police powers of the city government could intervene in the public's defense. There were thus great gaps in the control, whole sections of the city where physical as well as technological obsolescence ran its full course. Nevertheless, the building laws became an important and effective part of the control apparatus.

Aside from its impact upon the whole structure of the city, the automobile led to the development of a series of specialized building types, traffic forms, and recreational areas. This evolution was rapid. A short thirty years separated the county pike from the great four-lane, crossing-free, landscaped state highway; the old livery stable from the multiple-story garage; the general store from the suburban shopping center. Since Americans have from the start regarded the automobile as a prime source of entertainment and recreation, it is not surprising to find that, as the highway approached its present form, beauty as well as speed and comfort came to be considered legitimate criteria in highway design. Many of the highways were built to provide access to recreational areas, and it was not long before the highway itself began to merge with the park to which it led. The concept of the parkway appeared. The metropolitan area of New York may be said to have pioneered this type in the nineteen-twenties in the Bronx River Parkway and the magnificent Jones Beach development on Long Island. Here a whole system of new and existing recreational areas were linked by parkways into a homogeneous unit. A motorist could theoretically drive for days amidst their landscaped splendor.

THE HOUSING CRISIS IS DISCOVERED

With the entry of America into the First World War the housing problem was sharply raised on a national scale. The issue, in fact, was not new. As we have seen, the need for adequate housing for the industrial worker had been posed by the early New England textile factories. It had grown increasingly acute as the nineteenth century closed — not only for industrial workers but for all city dwellers. Thus there was already the beginnings of a housing movement. In addition to the growing awareness of the labor movement, there was an increasing body of professionals — social workers, municipal authorities, architects — who saw the urgency of the problem. With the war, the issue could no longer be dismissed as socialistic nonsense. In the Atlantic seaboard cities, housing conditions soon became an open scandal. The influx of workers into the war industries proved the inability of private enterprise to provide healthful housing at reasonable rates.

The crisis did not catch American architects wholly unprepared. English experience in housing and town planning, also accelerated by the war, had already attracted some attention on this side of the Atlantic. Although the Garden City was primarily a British solution to the problem of big-city growth, there was much in the physical planning that was applicable to any community. The emphasis upon open spaces, individual gardens, large blocks closed to vehicular traffic, appealed especially to American designers, caught in the sterile checkerboards of real estate expansion. A small but tenacious and eloquent group of town planners and architects — among them the elder Henry Wright, Clarence Stein, Frederick Biggers — finally succeeded in getting the government to undertake housing in some of the shipbuilding cities. In this effort they were mightily aided by Charles Harris Whitaker, then editor of the

Journal of the American Institute of Architects. To the manifest necessity of providing decent housing for the war workers, Whitaker added the weight of his own fiery pen and the writings of a whole circle of more experienced European housers.

The total number of housing units built by the two federal agencies was not impressive, and the abrupt end of the war laid both agencies open to the murderous fire of the vested interests. The agencies soon expired, the incompleted projects knocked down on the auction block, and the housers back on their own. But a beginning of inestimable importance had been made. Housing had at last been raised to the level of a permanent national issue, like the tariff or the sales tax. The architectural profession had been forced to accept housing as a legitimate part of its responsibility. The housing movement, in fact, was destined to become a veritable school for a whole generation of architects. For here, more than in any other building type, the architect was brought face to face with the basic social and economic forces at work beneath the surface of American building. It was not a pleasant picture; and the next decade was to furnish precious few examples of large-scale, low-cost housing. Wright and Stein were able to carry through their middle-class project at Sunnyside, Long Island, and their even more famous development of Radburn, N.J. — "Town for the Motor Age" (Figs. 190, 191). The Amalgamated Clothing Workers erected an apartment-house project in New York, the Rosenwalds subsidized a project for Negroes in Chicago. There was little else. But these few concrete examples were important out of all proportion to their size. They were widely published, criticized, admired. They had the salutary effect of shifting the architect's attention from esthetic abstractions to social reality. From the end of the First World War to the beginning of the Second, there was scarcely an architect who would not be to some extent involved in the housing movement — and better and wiser for it.

The new standards were to be largely formed by contemporary developments in European art and architecture although, in 1930, few American architects would have guessed it. In fact, their isolation during the first three decades of this century seems, in retro-

Fig. 190, 191. "Town for the Motor Age," Radburn, N. J., begun 1929. Henry Wright and Clarence S. Stein, architects and planners. First American response to English Garden-City Movement, Radburn was overtaken by the Depression, never completed according to original plan (below). Aerial view from mid-Fifties shows superiority of original Wright-Stein nucleus to subsequent speculative expansion.

spect, well-nigh incredible. After all, American painters had discovered modern European art in 1913;[10] American writers had discovered the Left Bank in Paris immediately after World War I; and young avant-garde European architects were already bringing these new ideas to America — Rudolph Schindler, Richard Neutra, Antonin Raymond, Paul Grotz, William Lescaze, Frederick Kiesler were all working in this country before the Great October Crash of 1929. Yet it was not until the thirties that American architecture began to respond to the revolutionary developments in Europe. This cultural lag is not easy to explain, for American architects of that period were, if anything, better traveled and more often educated abroad than other middle-class professionals. Part of the explanation may have been the unchallenged prestige of the Ecole des Beaux Arts in Paris. Vienna, Berlin and the Lowlands — scenes of such prewar stylistic movements as *Art Nouveau, Sezessionsstil* and *Jugendstil* and the postwar *De Stijl* and Bauhaus — were off the beaten track for young Americans during those years. And the Beaux Arts, which seemed to be as completely untouched by these winds of change as if it had been on the isle of Madagascar, had the curricula of American schools firmly chained to its eclectic chariot. Whatever the detailed reasons, American architects seem to have gone blindfold through the Grand Tour. If they saw or heard anything of the new architecture in Vienna, Berlin, or Amsterdam, it is not apparent in their work. Nor were those who stayed at home informed by their architectural journals: these, in the decade between 1919 and 1929, carried little news from the European front and grossly misjudged the significance of what they did report.[11]

This smug isolation was shattered by a now-famous event — a show of contemporary European architecture opened by the Museum of Modern Art in New York on February 10, 1932. Called *The International Style* by its organizers, Henry-Russell Hitchcock and Philip C. Johnson, it created an immediate sensation, very much like the historic Armory Show of 1913. Together with its catalogue and the book which subsequently grew out of it,[12] this show triggered the blast that lifted American architecture out of the parochialism into which it had slipped. It had not, of course, occurred in a vacuum: opening in a world wracked by depression, within a year of the Reichstag Fire and the inauguration of Franklin D. Roosevelt, and presented to a profession which was about 85 percent unemployed, the International Style raised issues which were already explosive. It served to restore to the agenda of architectural controversy those very questions originally posed by Sullivan and seconded by Wright but successfully suffocated by the Establishment for forty years.

For the more percipient of younger American architects, the show afforded a stunning confirmation of what they had begun, intuitively, to understand — i.e., that a completely new language of architectural form, consonant with the facts of modern life, was not only mandatory but *possible.* Escape from the prison house of historical eclecticism was a demonstrated possibility. Such a process can be traced with diagrammatic clarity in the work of the Philadelphia architect, George Howe. Junior member of one of the most fashionable firms in the East, Howe had hitherto been identified with the design of elaborate "period" houses along the Philadelphia Main Line. Yet when Howe read his paper, "Why I Became a Functionalist,"[13] at the Modern Museum during that same 1932 show, he had already been for several years hard at work on a basic philosophical re-examination of his own Beaux Arts training. And he and William Lescaze were even then completing the new skyscraper for the Philadelphia Society For Savings — a structure of germinal significance which William Jordy has called "a pre-eminent examplar of the abrupt appearance of the International Style in the United States."[14]

Searching for an "architectural expression which should not be in conflict with any form of modern activity," Howe said, "I felt I had failed either to evolve or discover such an expression until I became conscious of the meaning of the so-called modern system to the west in America and to the east in Europe."[15] The "west" to which he referred was the Southern California in which Wright was flowering for a second time; in which Schindler had years before finished his prescient beach house at La Jolla; and in which Neutra had more recently completed his sensational "Health House" for the Lovell family. The "east" to which Howe referred would include the Germany in which the Bauhaus was already exercising world-wide in-

Fig. 192. P.S.F.S. Building (above), Philadelphia, Pa., 1932. George Howe and William Lescaze, architects. Mezzanine Banking Room, P.S.F.S. (below). A highly successful demonstration of the esthetic potentials of skeletal frame and non-load-bearing curtain wall, the P.S.F.S. was the first building to return to the line of development begun in Chicago by the Reliance and the Schlesinger Buildings (Figs. 164, 165).

fluence; the Paris in which Le Corbusier was already recognized as an international force; the Holland of Dudok and Oud; the Scandinavia of Asplund. As a wealthy and well-traveled man, George Howe was an exception: he had taken the trouble to seek out these "many other men (who were) seeking a technically and expressive solution of modern architectural problems."[16]

The first opportunity that other American architects had to demonstrate that they too began to feel the new winds blowing from the East came, the following year and with rare poetic justice, in Chicago. There, in 1933, The Century of Progress Exposition opened its gates to a bankrupt nation which, as FDR had said, had nothing to fear but fear itself. The architectural message of this exposition marked the close of the eclectic interregnum which its predecessor, the World's Columbian Exposition, had proclaimed in 1893. That proclamation had marked the death knell of Louis Sullivan's hopes for a new American architecture. Sullivan was ten years dead when the Century of Progress opened: history seldom moves so quickly and so tidily to vindicate its victims.

Of course, while the propagandistic impact of that architecture can scarcely be overestimated, one must be careful not to exaggerate the intrinsic value of the buildings themselves. In retrospect, they seem pallid and superficial, inexpert and clumsy reflections of the new architectural doctrine. It is not that they were cheaply and shoddily built — that has always been true of exposition buildings and has never affected their seminal power. It is rather that not one of them was truly illuminated by any real comprehension of the *principles* that underlay such electrifying designs as those for Le Corbusier's Esprit Nouveau Pavilion in Paris in 1925 or Mies van der Rohe's Barcelona Pavilion (1929). They were not masterpieces, these first American essays in the new language: but then neither was *Uncle Tom's Cabin* a masterpiece. Historical necessity does not always have at hand the perfect vehicle for its expression; the buildings in Chicago became, willy-nilly, the instrument of change.

The Fair elicited mild praise and stern invective from the professionals. Conservative opinion, naturally, was dismayed by an architecture it felt to be "harsh," "inhuman," "communistic," "barn-

Fig. 193. Aerial view, Century of Progress Exposition, Chicago, Ill., 1933.

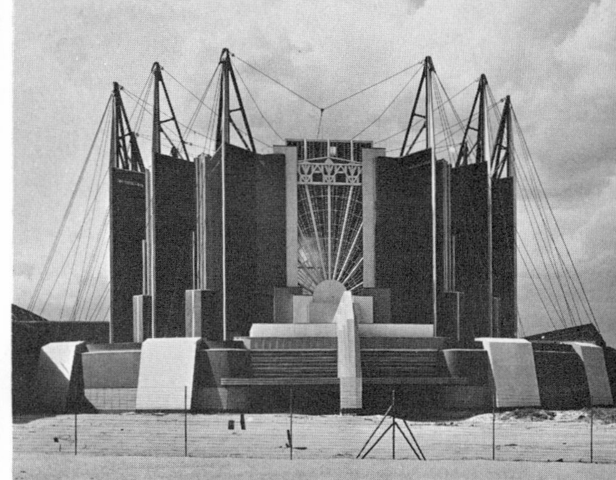

Fig. 194. Century of Progress Exposition: Federal Building (left) and Transportation Building (right), Bennett, Burnham and Holabird, architects.

like," "angular," "machine-for-living." The laymen who crowded into it responded with interest if not admiration. An analysis of the Fair issued by the Book and Print Guild of nearby Winnetka probably voiced public opinion accurately when it said that "as buildings, these of a Century of Progress Exposition are different and somewhat peculiar. They are even bizarre and strange. They are unlike any of those things in the world, so far, commonly recognized and known as buildings."

Nevertheless, the author had gone to some pains to try to understand the philosophy behind them. They exemplified, he said, "the theory of realism current in these times. This is the theory that beauty is the bare truth. Beauty is not a thing that is merely put on." And then he proceeds to prophesy:

> First, that the dignity and primacy of industry and commerce in the future will find honor and beauty in its own forms. The functional mysteries of mechanical and chemical production will bring forth structures that may be strange and even bizarre. This is the real Modernism.
>
> The second is that the architects, the designers of the more institutional buildings of the future will no longer return to their copy-books and the conceits and motifs of the dead past. And it is to be hoped that they will not take up the premature orthodoxy of German and French Modernism![17]

Another event, ultimately of great significance in the history of American taste, had occurred even before the opening of the Century of Progress. This was the beginning of the restoration of Williamsburg, Virginia, by John D. Rockefeller in 1929. Acting in exactly a contrary direction to the Chicago event, this project was to do as much to stultify popular taste as any single development in our history. An ironic outcome, inasmuch as the restoration itself has always been handled with immaculate taste and its reconstructions of eighteenth century buildings has always been in the hands of skilled and knowledgeable specialists who seldom acted without adequate research and documentation.

The negative impact of "Colonial Williamsburg" has been largely due to the bowdlerized version of American life which, perhaps unwittingly, it has propagated. This distortion stemmed from two awkward technical problems common to any museum of this type. The first is that time has been telescoped: buildings which never coexisted at precisely the same point in time and space have been restored to an artificial simultaneity. The resulting image is one of polished and manicured amenity which is not so much untruthful as incomplete. It is nonetheless misleading. The second problem is related to the first. Williamsburg in pre-Revolutionary times was a small provincial capital, technically backward and riddled with class divisions including human slavery. It is difficult to "restore" this aspect of social reality. Slave pens and muddy streets, hunger and discomfort are difficult to display museologically: the natural tendency of any curatorial staff is to select, dramatize, hence prettify. Thus the result has been the creation of a stream of beguiling half-truths about the American past, all too easily vulgarized and fed into the bloodstream of popular taste. One result has been the appearance of "Williamsburg architecture" in places, climates and circumstances where it was grotesquely misplaced. On the other hand, with the passage of time, Williamsburg has become a center of archeological research and curatorial experience in the conservation of works of architecture and art. Perhaps, at a higher level it has thus atoned for damage done at a lower one.

Fig. 195. The Governor's Palace, Williamsburg, Va.; built 1706–20, restored 1934.

8. 1933–1945

AMERICAN BUILDING AT THE CROSSROADS

By the time of the second Chicago Fair in 1933, the crisis in American building had become chronic. Every phase of the field was involved: concept, architect, design, and actual building. The crisis had been maturing throughout the three-quarters of a century which had elapsed since the Civil War and had been marked throughout by this curious fact: though building had been subject to many invasions by science, it had never been conquered by science. Even in the forepart of the current century, when the whole field had been rocking under the impact of technological advance and scientific discovery, building had managed somehow to preserve its profoundly anti-scientific conceptual basis. It had managed to absorb the elevator, the steel skeleton, the electric power line, without really absorbing the concepts, the philosophies, and the disciplines which made these developments possible.

Indeed, as often as not the building field had displayed an actual hostility toward science and technology, and this tendency had become very sharp among — of all people — a section of the architects themselves. In increasing numbers, these men had seen the main task of building handed over to the engineers. The struggle to meet the needs of their society had proved too great for these tender spirits, and for solace they had turned increasingly to the past. These were the men who, like the critic A. Kingsley Porter, could see only that "good architecture came to a sudden end in America about the year 1850. . . . It was the machine which crushed out handiwork, it was the machine which killed beauty."[1]

Not only architecture but human nature itself had changed for the worse. Thus the successful architect Thomas Hastings saw the medieval mason as a man who "praised God with every chisel stroke . . . for him work was worship; and his life was one continuous psalm of praise. . . . Now a Gothic church is built by laborers whose one interest is to increase their wages and diminish their working hours."[2]

There were many of scholarly temperament who accepted this analysis, but there were other conservatives who, while admitting that things were bad, still were unwilling to say that all was lost. Typical of this sort of reaction to the inexorable pressure of science and technology upon American building was the analysis of William Adams Delano. He had come to the conclusion that "building is a business. Architecture is an art . . . to draw a distinction between building and architecture is [however] not easy, it is to be felt rather than described."[3]

Here Mr. Delano, senior member of one of the country's most fashionable firms, was trying to draw a charmed circle of "art" around a certain portion of building — the very top — in order to protect it from the ravages of historic fact. Unfortunately, Mr. Delano's line of exemption (which he admitted was to be felt rather than described) did not in fact exist. The contradictions raised by Mr. Porter's machine cut across the whole fabric of American building, the clubs of the rich as well as the slums of the poor.

There was, finally, that bedrock of Tory opinion which blindly and stubbornly refused to admit even the possibility that preindustrial concepts of plan and façade were not entirely adequate to twentieth-century needs. This school was typified by John Russell Pope, whose last two commissions — the Jefferson Memorial and the National Gallery of Art in Washington — are a perfect reflection of this concept. The very existence of such structures, it might be added, is sufficient proof that Mr. Pope moved in no vacuum, but spoke for a powerful laity whose concepts he accurately reflected (Figs. 196, 197).

Fortunately for the reputation of American architects, there had been another and less tarnished side to the coin. For architecture had scarcely achieved its modern status as an independent profession in the mid-seventies when voices within it began to be raised in defense of science and technology, and demands made that architectural concepts be forced into consonance with that fact. It was, of course, an immensely difficult problem, then as now, for the full implications of the Industrial Revolution were not apparent. Thus, in different men, according to their temperament and experience, the issue was expressed in different forms and analyzed at different levels.

Louis Sullivan, who had seemed at one time so near to formulating an effective concept, had seen it as both a political and an esthetic imperative. A democratic way of life would produce a democratic architecture; satisfactory function would produce satisfactory form. His thunderous eloquence was echoed in the architectural press by men like Montgomery Schuyler, A. D. F. Hamlin, and Claude Bragdon. For Frank Lloyd Wright the issue had been at once esthetic and philosophical — he wanted organic unity restored to both architecture and life. To planners like the elder Henry Wright and Clarence Stein the issue seemed largely technical and the solution physical — one had only to organize the land intelligently. Meanwhile, to their contemporary, Charles Harris Whitaker, a decent democratic architecture was an ethical if not a moral necessity.[4] Lewis Mumford had expressed the concept in terms of man's right to mental and physical well-being. Technicians like Buckminster Fuller were baldly mechanistic — the problem of twentieth-century building was simply one of applying modern technique. A house was, as Le Corbusier had so famously put it, merely a machine built for living in. Around such spokesmen as these, there had grown up a whole generation of social housers, planners, and economists. For them, the issue was variously seen as one of clearing our slums, decentralizing the cities, or lowering the cost of building.

These were all attempts, by the building professionals themselves, to analyze the crisis, to formulate new concepts which would enable them to extricate building from the slough into which it was slipping. None of them was complete, yet none of them was wholly fallacious; and taken together, they did constitute the basis for the genuine integration of building with science and technology. Together, they were a many-faceted formula for the resolution of the cruel dichot-

Fig. 196. Jefferson Memorial, Washington, D. C., 1939. Office of John Russell Pope, architect. The last important monuments of the last wave of eclectic revivalism. The Jefferson Memorial and the National Gallery of Art (below) perfectly confirmed Sullivan's prophecy that the reactionary impact of the Columbian Exposition would "endure for half a century."

Fig. 197. The National Gallery, Washington, D. C., 1941. Office of John Russell Pope, architect.

omy between technique and esthetic, between bountiful promise and poverty-stricken reality. Such concepts had been seventy years agrowing, but it was within the twelve years between Franklin Roosevelt's inauguration and his death at the close of the Second World War that they matured. For the building designer, the period was indeed a stern if fruitful school. It opened with as many as 85 per cent of all architects and engineers unemployed. It closed with fully that percentage either in the armed services or exclusively employed in military construction. The years between had seen our society faced with problems of unprecedented scale and complexity — problems which involved the architect in every level and sphere of his existence. It was in such an atmosphere that these new concepts of the function of building spread, deepened, intertwined, to emerge finally as the recognizable body of theory and practice known today as modern architecture. If the period produced more ideas than buildings, both were pregnant with change. If the buildings of the thirties were statistically less important than those of the twenties, they were nonetheless far more significant. For the concept which they expressed was catalytic — namely, that the function of American building must be the maintenance of those optimal environmental conditions essential to the health and happiness of the individual and to the peaceful and efficient development of American society.

CONFLUENT TENDENCIES

Of the several broad tendencies which flowed together to form that body of theory and practice known as modern architecture, the first and most characteristic was that of native common sense. In the eighteenth and nineteenth centuries this led to a pragmatic functionalism that found eloquent expression in the anonymous popular building of the carpenter, the "mechanick," and his descendant, the engineer. It had had a theoretical apparatus of a sort, too, in the handbooks and textbooks which came off the presses in increasing numbers. But, being largely self-trained, these men had no academic or professional ambitions; and though their point of view found occasional literary expression in such men as Greenough and Emerson, it had little or no impact upon official architecture. It merely formed, as we have seen, the huge submerged base on which the architectural Establishment erected its eclectic monuments.

In the present century, however, the most technically advanced sectors of business, industry, and commerce began to demand a more functionally efficient architecture. And, once their demands were met on a purely practical plane, it was inevitable that their esthetic consequences would be expressed at a higher, more abstract level. Simplicity, economy, and efficiency became formal, as well as functional, criteria.[5] A whole body of work appeared in which this common denominator underlay the vagaries of individual architects and clients. While this movement was large and not always distinctly separated from other tendencies around it, the essential character can be circumscribed with a handful of names: the long series of industrial plants designed by the office of Albert Kahn; the skyscrapers for the *Daily News* and the McGraw-Hill Publishing Company in New York designed by Raymond Hood; the group of buildings at New York's Rockefeller Center — joint work of three firms, including Hood's;[6] the General Motors Exhibit by Norman Bel Geddes and the Ford Motor Company Exhibit by Walter Teague — both at the 1939 New York World's Fair. These structures reveal one striking characteristic in common — even beyond the necessities of structure and plan, they depend chiefly upon simplification for effect.

Stylistic simplification was the expression on the esthetic plane of the underlying pressure for economy all during the thirties. Economy of initial cost, economy of maintenance, economy of space, manpower, and mechanical energy. In its crassest form, economy meant simply cheapness. A landlord wanted the most for his money — the largest area possible in this apartment or that factory for the least possible cost. Under such a set of pressures, something in conventional building had to give way. Rationalization of construction methods yielded some cost reductions. A certain amount of cheapening or adulteration was possible in the structure proper —although this was limited by the building codes. The rest of the paring had to come out of the fat of the building — its ornamental or decorative features. There had to be less of them, or they had to be cheaper, or both. In the early part of the century, this necessity had led to a flourishing business in "mail-order Classic" — plaster and terra-cotta ornament stamped out by the thousands and sold by the yard. These were cheaper than carved wood or stone details, but — after the crash of 1929 — not cheap enough. Ornaments as such had to go. For a brief transitional period, the old devices hung on the pilasters, flutings, and swags reduced to nothing more than etched lines on the skin of the structure. Finally they disappeared altogether.

Building design gained from their disappearance, even though the improvement was superficial and the essential mass and articulation of such structures remained unchanged. For the architects themselves, the process was a blessing in disguise. It was the first time that many of them had been forced to face the problem of genuinely economical structure; and, in the solution of such problems, it was inevitable that the principles thus evolved would find expression in new esthetic concepts.

Simplification could spring from more valid motives than mere penny-pinching, however; and, as expressed in the outstanding buildings of this group, it could lead to more significant results than merely stripping the façade of useless ornament. It could and did produce hundreds of buildings which were handsome in quite a new fashion — buildings whose appearance of orderly precise articulation was clearly the result of a radically different approach to the design process. Thus, in many of the Kahn factories there is a very high esthetic standard which appears as an almost incidental by-product of the main work at hand — a pragmatic solution to a set of very practical problems. Appearance and purpose, function and form, are so intimately related that it is difficult to separate the two. One might almost imagine that any formulation of their esthetic principles must have *followed* the actual invention and development of the idiom (Fig. 200).

Yet in the *Daily News* or Rockefeller Center Buildings, the same esthetics are at work as a conscious principle, deliberately applied. Here in visual terms is a statement of the power of simplification. The reference to industrial production is inescapable. In the factory it leads to impersonal, continual, and precise multiplication. In the skyscrapers it produces the cold and polished surface, the endless

Fig. 198. Daily News Building, New York, N. Y., 1930. Raymond Hood, architect.

Fig. 199. McGraw-Hill Building, New York, N. Y., 1930. Raymond Hood, architect.

Fig. 200. Chrysler Truck Plant, Detroit, Mich., 1937. Albert Kahn, Inc., architects.

soaring line, the meticulous geometry of intersecting column and girder. At the two Fair exhibits, the same principles were at work in even more explicit terms. The curve — so studiously avoided in the early thirties because of its connotations of the masonry arch — reappeared as a symbol of smooth, mechanical motion. The curving lines of the Teague and Bel Geddes designs (Figs. 201, 202) by implication register the immense advance of a short decade in understanding of flow lines, whether of stresses in a steel member or traffic on a factory floor. Teague's use of the spiral ramps with their moving, multi-colored cars thus evoked both the smooth highways on which they glided and the assembly lines on which they were born. In Bel Geddes' design for General Motors the right angle disappeared altogether. The very forms were reminiscent of the streamlined vehicle. Though the stucco skin belied it, the structure was externally modeled into shapes which recall those of a diesel locomotive or a motorcar body. It was a totem to industrial production.

For all their handsomeness, the buildings of this group were often surprisingly cold and mechanistic. Almost without exception they were severe — in their more formal aspects, even autocratic. Stylistic simplification could itself become a fad: thus the continuous vertical piers of the Rockefeller skyscrapers were only a partial statement of the structural problem. The gross overemphasis given them was itself dishonest. Clearly, simplicity and efficiency were not a broad enough base upon which to erect the esthetic of modern architecture.

There was a second current within the modern movement whose sources can be plotted in the works of Richardson, Sullivan, and Wright. It represented another and equally persistent set of American traits: the idealist's demand for a way of life in which artistic imagination and intuition could play upon and fertilize the bare facts of material existence. There was only a small segment of the building field in which this tendency could express itself, however: upper-class residential work, an occasional church, an even rarer public building or tomb. Business, industry, and commerce were generally opposed to poetic overtones for both practical and philosophical reasons. And — disregarding the personal factors of background,

Fig. 201. General Motors Exhibit Building, New York World's Fair, 1939. Norman Bel Geddes, designer.

Fig. 202. Ford Motor Exhibit Building, New York World's Fair, 1939. Walter Dorwin Teague, designer.

temperament, and opportunity which gave them the direction they took — it is apparent that the large and stormy individualism of men like Wright and Sullivan made them ill-fitted for the life of a successful corporation architect. This was fortunate for modern architecture, for domestic design permitted more esthetic experimentation than any other segment of the field.

The ravishing perspectives of Wright's own houses at Taliesen East and Taliesen West may well serve as index to the ideals of this current of American design. The exotic forms and magnificent compositions; the artless yet cunning integration of indoors and out, landscape and structure, warm colors and sensual textures — these are the elements of an esthetic just as indigenous as Albert Kahn's, but very different in intent. As Lewis Mumford so succinctly put it at the time, Wright "depolarized regionalism from its connection with the historic and the archaic: he oriented it towards the living present."[7] His houses had none of the eclectic's dependence upon the past. Where allusion did occur it was oblique, abstracted. Here and there in the Wrightian landscape one might catch a fleeting color from the American Indians, a glint of burnished copper and polished wood from some Quaker kitchen, a shingled eave which evoked the frontier cabins of Audubon. But these images were always abstracted: as in all great architecture, the observer was never quite sure whether they existed in the building or in his own head.

But for one authentic poet like Wright, the romantic current of American architecture had always had dozens of storytelling hacks. The eclectics of the late nineteenth and early twentieth centuries had erred, not in wanting to introduce poetry and romance into modern life but in relying upon literary means (and secondhand means, at that) to accomplish it. Thus every French villa in Newport or Spanish palace in Palm Beach, every Roman courthouse or Renaissance banking room, was a literary anecdote, not an artistic invention. And no amount of Beaux Arts training, no amount of virtuosity could long conceal the fact. "They try to make stories of everything," Wright had once sarcastically exclaimed, "painting, music, architecture!"[8]

Wright's rediscovery by a younger generation, when he was already in his sixties, was spectacularly marked by the great one-man

Fig. 203. Living Room, Taliesen, Spring Green, Wis. Frank Lloyd Wright, architect. Though he was exposed to many cultural stimuli, Wright always transformed them into forms inescapably his own.

Fig. 204. Stair Hall, Mrs. Hugh Dillman house, Palm Beach, Fla., 1926. Addison Mizner, designer. Reproductions of Spanish and Venetian furnishings filled fake Spanish palaces.

show which the Museum of Modern Art opened in November, 1940. Strictly speaking, he had produced no disciples. But his work had created a climate of thought and feeling which nourished a whole generation of younger men just coming into practice in the decade before the war. Whatever their specific origins, men like Alden Dow, O'Neil Ford, and Harris Armstrong in the Midwest and Harwell Harris, William Wurster, and Gardner Dailey in San Francisco obviously belonged to the Wrightian phylum of design. The houses they built during these decades reflected, like Wright's, the best aspects of American domestic life — its humanity, hospitality, and good sense.

In view of their restricted area of operations, the influence of this group upon the whole body of architectural thought has been remarkable. It must be remembered that their dominance in residential design was not won without a long and bitter struggle. Expensive residence work had long been considered the preserve of a whole group of highly skilled eclectics: the famous firms of Delano and Aldrich; Mellor, Meigs and Howe; Addison Mizner; Frank Forster and Harrie Lindeberg. These men were also imaginative, skilled at composition, texture, and color. For decades they had been the unchallenged ideologues of high style, spokesmen for the wealthiest families in America. Their privileged position was destroyed by the depression of 1929–33. An abrupt end was brought to the succession of great houses which had dominated the domestic scene for so long. At the same time, the ideals and perspectives of the class which built them were suspect, discredited (Fig. 204).

Yet another group, the technologists, were exerting great influence on the building field, even if in indirect fashion. Their most persuasive spokesmen were the engineers and the newly created profession of industrial designers. These men were busily designing locomotives and airplanes: they were advocating that architects adopt not only the processes of industrial technology but its vocabulary of form as well. Buckminster Fuller was a leading theoretician for this current of opinion. Although himself a well-educated and polished man, he belonged much more to the tradition of the great self-made pragmatists like Roebling, Paxton, and Edison than to the genteel tradition. He produced during this period a series of provocative

Fig. 205. A. B. Dow house, Midland, Mich., 1935–41. Alden B. Dow, architect.

Fig. 206. H. H. Harris house, Los Angeles, Calif., 1939. Harwell Hamilton Harris, architect.

designs for a streamlined three-wheeled automobile; for a completely prefabricated bathroom; for a self-contained industrialized house hung from a single mast. But, though they achieved immense publicity, none of them got beyond the mock-up stage. Fuller's prewar success was real but propagandistic: none of his projects were to get into actual production until after World War II.

These were the native forces at work in American design in the years before the war. Even by themselves, they would have radically altered its physiognomy. But this process was enormously accelerated by the arrival here of a second wave of European architects. By the war's end, many of Europe's most famous practitioners were in the United States: Alvar Aalto, Curt Behrendt, Marcel Breuer, Serge Chermayeff, Walter Gropius, Eric Mendelsohn, Ladislas Moholy-Nagy, Mies van der Rohe, José Sert, Konrad Wachsman, Roland Wank. Coming from practically every country in Europe, these men brought with them a rich and varied experience: stylistically, they represented a complete spectrum of avant-garde opinion. Their influence was to be very great, the more so that they were well trained in polemical discussion and many were to become successful teachers.

Maturation was due most of all to the three giants: Gropius and van der Rohe (both of whom assumed control of important architectural schools) and Le Corbusier (who, of course, remained in France). The fructifying impact of these men upon our theory and practice is literally beyond reckoning. Nourished and disciplined as they were by the much more rigorous intellectual climate of Europe, they introduced us to formal, structured, and coherent theories of architecture such as we had never known. And it was one of the happier accidents of history that their theories complemented each other almost perfectly. Together with Wright they represented in talent and philosophy just about every possible approach to the problems of the field.

Initially, Gropius was perhaps the most influential. As the founder of the world-famous Bauhaus, his philosophy of design education was most relevant for the United States. For ours was the most thoroughly industrialized nation and the most acutely wracked by

Fig. 207. First "Dymaxion House," 1927. R. Buckminster Fuller's design for a completely prefabricated, factory-produced house. Entire house was hung from central mast which also enclosed all utilities.

Fig. 208. Second "Dymaxion House: 1947." After World War II, it was widely assumed that the swollen aviation industry would enter the housing field, prefabricating units like this improved design by Fuller.

Fig. 209. "Dymaxion Car: No. 3," 1933. Fuller's design was based on complete, authentic "streamlining." It also included many safety features not yet incorporated into the Detroit-made auto.

social, ethical, and esthetic problems stemming from that industrialization. Gropius's theories are not easily compressed but three salient points were these:

1. that industrial production must necessarily be the basis of the modern esthetic, just as handicraft production had been the basis of preindustrial standards of beauty.

2. that cooperative work between design specialists would be as essential in the design studio as on the production line.

3. that to produce good modern design, the schism between head and hand, theory and practice, designer and worker, would have to be healed by a unitary theory of education. Although employing a visual syntax quite different from Sullivan's, the Bauhaus had developed a rationale very similar to Sullivan's thesis *Form follows function* — i.e., valid form derives only from function understood and served. The viability of this rationale was proved then (and proved now) by the durability of the designs it produced — buildings, china, automobiles, furniture — which, after four decades, are still in fashion.[9]

Mies van der Rohe's influence stemmed from a different base. While no less rigorously principled than the others, his canons of design were those of the meticulous craftsman (he had been apprenticed as both marble worker and cabinetmaker) rather than the theoretician. His lessons were taught visually, not verbally. When he came to Chicago in 1938, he had written few essays or taught few classes; but he had delivered two of the most powerful sermons in the modern world when he built the Barcelona pavilion and the Tugendhat house. Mies's commitment, then as now, was to a truly Platonic perfection of form. He was never a functionalist like Sullivan nor an "organicist" like Wright. Yet such was his genius that he could take over the steel-framed skyscraper, where it had been left fifty years before, in the Schlesinger and Meyer department store and the Luxfer Prism Building, and carry it on to new heights of formal elegance. This skin-and-bones tradition in Chicago had always had a kind of single-minded directness, a naïve candor that was sometimes admirable, more often merely boorish. Mies, with his merciless perfectionism, was to raise the idiom to astonishing

Fig. 210. The Bauhaus, Dessau, Germany, 1925. Walter Gropius, architect. Like Jefferson at Charlottesville, Gropius was architect of both campus and curriculum of this world-famous institution.

Fig. 211. German Pavilion, Barcelona International Exposition, 1929. Ludwig Mies van der Rohe, architect. Perhaps the single most influential building of the inter-war period, its fame rested wholly on photographs: it stood for only one summer and was seen by very few architects or students.

levels. Like Gropius, he aspired to a suprapersonal system of expression: even better than Gropius, he succeeded in teaching it to the first postwar generation. The system had serious functional limitations (as I have pointed out elsewhere[10]) but for the immaculate consistency with which every element was developed, it had no equal in the world.

The third great vector in this complex of European forces was Le Corbusier. He was in many ways the most subtle of them all, for he combined a poetic imagination of great range and power with a truly philosophic approach to social reality. He had, moreover, the classic French capacity at abstraction. These twin capacities, so rare in art, have always been present in Le Corbusier though the balance between the two has shifted from time to time. It happens that, during the thirties, he was still immersed in the search for ideal solutions of socially urgent problems: the fantastic linear project for Algiers; the Paris headquarters of the Salvation Army; the project for the Palace of the Soviets — all these from 1931; a proposed university city for Rio de Janeiro (1936); the "Growing Museum" of 1938. All these designs were models of coherence and rationality, in their optimism and lucidity deriving from the tradition of Saint-Simon and Fourier. Yet he was not neglecting the task of evolving an adequate visual language for the celebration of these social goals. This is clear for example, in his moving design for a memorial to his friend Vaillant Couturier.

Le Corbusier's creations instantly became what Giedion calls "constituent facts" of the International Style, studied and mastered or mimicked everywhere. One has only to compare them with urbanistic work around the world to see how literally true it is that Le Corbusier, like Picasso, is one of the actual creators of the modern vision. However, in those prewar years, he was perhaps less influential in the United States than in South America and Asia. This was not unconnected to the fact that the special plastic qualities of his designs depend upon his favorite material, reinforced concrete. And American designers, for all their wide use of this material in heavy engineering, had been peculiarly loath to experiment with it as an "architectural" material. Of course, our prejudice in favor of skeletal construction and linear designs derived from our abundant supplies of suitable materials — first timber, then structural steel. (Wright was almost the only architect who moved freely between linear and mass designs.) It was only after World War II that our use of concrete was to approach European practice; and a number of young engineers like Mario Salvadori and Paul Weidlinger, trained in the more daring European tradition, contributed greatly to this new confidence. This, in turn, made possible a new appreciation of Le Corbusier.

But he, in the interval, had also changed. As it became clear to him that the organs of official power (the French Republic, the United Nations) were not disposed to entrust him with the execution of those projects of social reconstruction so dear to him, the balance between the practical and poetic began to shift. This shift is clear in the church at Ronchamp and the monastery of Le Tourette (Figs. 216, 217). Here he has given full rein to the irrational component of his imagination with results that are majestic, somber, literally awe-inspiring. This same quality — muted naturally by practical requirements — informs the great brooding silhouettes at Chandigar. The power of this new monumental style of Le Corbusier is overwhelming. Little wonder, then, that it was profoundly to affect the postwar work of such Americans as Louis Kahn and Paul Rudolph.

Fig. 212. "Voisin Plan" for the redevelopment of central Paris, 1925. Le Corbusier, architect. This design for a "sky-scraper-studded park" was to be the prototype for American urban redevelopment.

The attack on Pearl Harbor brought an abrupt end to civilian construction. An enormous industrial and military construction program began. And though some of these wartime projects achieved a certain architectural distinction, it was generally speaking a period of enforced creative idleness. The experiences since 1933 could be viewed in retrospect, its accomplishments analyzed for their significance. For so large and rich a nation they had not been very impressive. The most significant had been supra-architectural in scale — the Tennessee Valley Authority, the Greenbelt towns, the public housing projects in many cities. And though they were admirable from an urban or regional point of view, they were usually unexceptional architecturally. The real work of theoretical and visual exploration had been done in projects of small scale and minor importance, many of which were never built at all. But the hiatus of the war years offered the entire public, lay and professional alike, an incomparable opportunity to review and digest the avant-garde lessons of the thirties. Thus it was clear, from the moment that controls on civilian construction were lifted, that the new architecture to which so many had contributed so much had emerged the undisputed victor in the field.

Fig. 213. Sketches for Ministry of Health and Education, Rio de Janeiro, 1936. Le Corbusier's style of delineation became immensely influential, displacing the Beaux Arts drawing style.

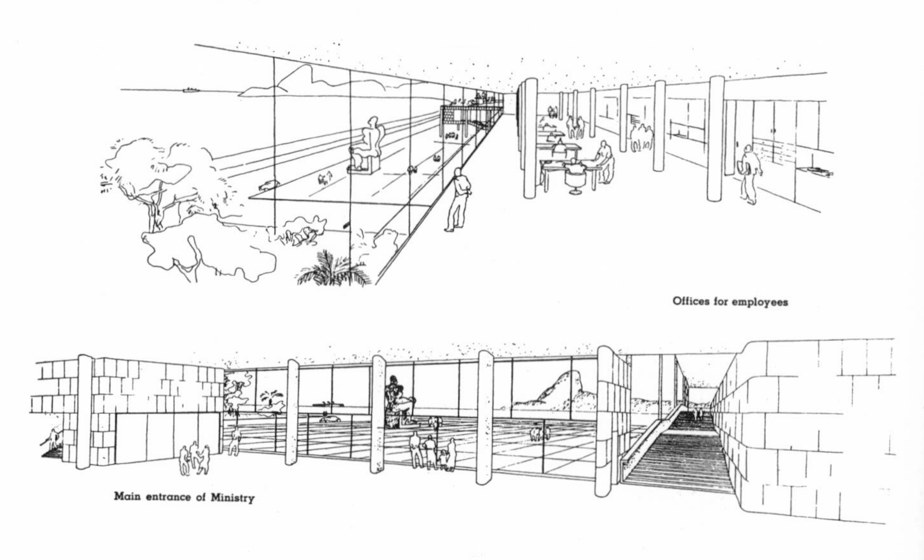

9. 1945–1965

THE PARADOX OF ABUNDANCE

Just as the First World War had afforded Europe a germination period for the avant-garde theories of art and architecture which were to flower in the twenties, so too did World War II prepare the way for the triumph of contemporary architecture in America during the two decades which followed it. The style emerged as the dominant idiom among the architects themselves: and it found a public educated during all the years of war to accept it. For the enforced idleness of the building field had provided a period for speculation, discussion, and assimilation of the various contending theoretical approaches to architecture. This was, in itself, a rare experience for Americans who have traditionally been "too busy to think" and it was reflected in the relative unity of taste in the first postwar years.

One could say, in all justice, that this taste and the style of expression it employed had been the joint creation of the four great makers — Wright, Gropius, Le Corbusier and Mies van der Rohe. For it is perfectly clear, in retrospect, that the lifework of these four dovetailed almost perfectly. In retrospect, it is apparent that the corpus of their works is almost uncannily complementary, each to the other — the poetic, anti-cerebral, and individualistic Wright; the poetic, intellectual, socially oriented Le Corbusier; the socially responsible, technologically conscious, and rational Gropius; the coolly detached and rational Mies. Taken together, the lifework of these four men covered — better still, *constituted* — the whole spectrum of esthetic possibilities of the first half of the present century. In a very real sense, it was they who had liberated the profession and the architectural schools from the prison house of Beaux Arts eclecticism.

This homogeniety of taste was soon to be fragmented, however, by the postwar boom. Peace released an uncontrollable flood of dammed-up civilian demands for new buildings of all sorts, in unprecedented quantities. The forced draft of war had resulted in an industrial technology qualitatively more sophisticated and quantitatively more productive than ever before. The result was an architectural profession more fully employed, and a buildings industry operating at more nearly full capacity, than had been seen since the boom days of the mid-twenties. But this prosperity was expressed at the ideological level by conflict and confusion.

The younger generation of architects, educated under the new freedoms won by the pioneers, now turned on the pioneers themselves. The maturation of this rebellion was marked by a symposium, at the Museum of Modern Art in New York, entitled *What Is Happening to Modern Architecture?*[1] The attack was against "functionalism": it was decried as mechanistic, restrictive of individual expression, too committed to sociopolitical (as opposed to artistic) goals, etc. Much the same charges, in other words, as had been raised against the modern movement a quarter of a century before. An attentive reading of the great pioneers might have absolved them of any such intentions, as Lewis Mumford tried to point out that night. They were jousting with windmills: what was being attacked as functionalism was a one-sided interpretation of function — "an interpretation that Louis Sullivan, who popularized the slogan 'form follows function,' never subscribed to."[2]

And Walter Gropius, at the same session, felt it necessary to repeat that

> functionalism for us meant embracing the psychological problems as well as the material ones . . . emphasis was not so much upon the machine itself as on the greater use of the machine in service for human life . . . Looking back, I think we dealt not too *much* with the machine but too *little*.[3]

But Gropius, as the first head of the Bauhaus, was to be subjected to increasingly bitter attack and ridicule in the following decade as the apostle of "Bauhausism" while Mies van der Rohe — paradoxically,

Fig. 214. Price Tower, Bartlesville, Okla., 1953–55. Frank Lloyd Wright, architect.

Fig. 215. Rotunda, Solomon R. Guggenheim Museum, New York, N. Y., 1959. Even in the tenth decade of his life, Wright repeatedly demonstrated the vigor and viability of his architectural approach.

the last head of the same institution — was to enjoy immense prestige with the same critics. Even as late as 1964, one leading historian was sarcastically to observe that all the "academic honors and medals [which] have so profusely acknowledged the Bauhaus doctrine of architectural education" were gratifying: "because never before has a curriculum turned out such a star roster of infidels: Johnson, Lundy, Barnes, Rudolph, Franzen, and others have revered their teacher while confounding his teaching."[4] It might have been more accurate, on the contrary, to have said that such a roster of pupils was precisely the validation of the Bauhaus "doctrine" — i.e., that (as Gropius never tired of saying) it aspired to teach a way of thinking, a methodology of approach, instead of a style.[5]

In any case, whatever their merits as designers, the ideological postures they assumed lacked the comprehensiveness, depth, and responsibility of the great pioneers. Most of them were celebrating a new kind of eclecticism while propounding various rationales of personal freedom which often sounded puerile, subjective and anti-rational. Ulrich Franzen is perhaps typical of the inversion which was overtaking many men of his generation, not only in architecture but also in painting, sculpture, and the theater.[6] "The great drawback to functionalist theory," he said in an interview explaining his changing philosophy of design, was that it "confirmed the idea that art, like everything else, must serve society and history — that art does not exist for its own sake. Modern architecture came to be seen as an ideology, a tool of technological progress." In explaining his "disenchantment with functionalism," Franzen explained that his liberation from its iron grasp was facilitated by his discovery that Wright, Le Corbusier, and Mies "simply did what they have always done, following the dictates of their own eyes and hearts. They used the fashionable rationalizations only long enough to give their work a cloak of respectability." This standing of history on its head seemed not to trouble either Franzen or his interviewer. Yet the theoretical formulations of these men were anything but "fashionable" when they were first propounded. And the suggestion that they were merely hunting "cloaks of respectability" is contradicted by the very facts of their lives. Each of them, by the strictest and sternest commitment to a hard-wrought set of principles, had

Fig. 216. Notre Dame du Haut, Ronchamp, France, 1951–55. Le Corbusier, architect.

Fig. 217. Ste. Marie de la Tourette, Evreux-sur-Arbesle, France, 1953–59. Le Corbusier, architect. In these buildings — the one lyrically plastic, the other sternly geometric — he created epochal images.

evolved a personal idiom of artistic expression in the face of bitter opposition. And the internal logic and consistency of their artistic and ideological development is amply established both by their buildings and their polemical defense of them.[7]

Even a casual reading of the American architectural press today will reveal that a very large number of men are at work upon a great range of projects. From such a survey it will be apparent that a wide range of problems, technical and esthetic, have been substantially solved. Yet this work, in its competence and novelty, may well give the impression that American architects are currently engaged in a more important task than is, in fact, the case. For these journalistic accounts tend to conceal as much as they reveal. First, they deal with individual buildings in preponderantly visual terms: and this inherently visual-esthetic bias leads to the building's being incorrectly or inadequately analyzed from the point of view of total environmental performance. Secondly, the journals do not — in fact, from the very nature of the medium, *cannot* — communicate much of an impression as to the actual quality of the landscape in which the building is literally imbedded. The net result of all this is a misleading account of our accomplishments.

For American architects, irrespective of their talents or philosophies, operate within the terms of reference established by American society. This imposes certain characteristic limitations upon their work which even the most talented or successful of them cannot rise above. In many respects, our architecture displays the highest material standard in the world. But American accomplishments must be measured against American potentials; and, from this vantage point, the record is much less reassuring. For example, compared with our accomplishments and ambitions in nuclear and space technology, our architecture and (especially) our city-building can only appear as old-fashioned, timid, eclectic, despite their avant-garde trappings. As long as architects were confined to the design of isolated vessels, it was still possible to analyze their work from a microscopic point of view — i.e., as discrete and independent works of art. And from this now-obsolete vantage point, the U.S. is still producing many handsome and civilized buildings. But the magni-

Fig. 218. 860 Lake Shore Drive, Chicago, Ill., 1951. Ludwig Mies van der Rohe, architect.

Fig. 219. Seagram Building, New York, N. Y., 1952. Ludwig Mies van der Rohe and Philip Johnson, architects.

Fig. 220. Dr. Edith Farnsworth house, Fox River, Ill. Ludwig Mies van der Rohe, architect. Another of the giants from the formative days of modern architecture, he still sets its parameters.

tude of our building activity today forces us to analyze it from a macroscopic — i.e., urbanistic — point of view. And in this context even our best work is less luxurious, less urbane, less truly effective than it ought to be: for at this level, the measure of architecture is not merely visual appearance but *total viability.*

The most disturbing aspect of life in the United States today is the widening discrepancy between privatized luxury and public amenity. The postwar years have witnessed a relative decline in public facilities of all sorts. This is expressed both in our failure to construct enough new schools, hospitals and public housing to keep abreast of current need and in the falling levels of maintenance of public transportation, parks and playgrounds, water and sanitation systems. This decline is not registered by the picture magazines because it involves so many nonvisual aspects of existence. It goes largely unremarked by editors, critics, and opinion-makers generally who, as middle-class intelligentsia, are themselves largely insulated against discomfort and inconvenience by the private amenities which surround them. Education, prestige, and economic status permit them to leap over the rising tide of inefficiency and malfunction which characterizes life for most Americans today.

Nor does all the architectural and urbanistic activity of recent years seem able to halt this tendency. It often appears, on the contrary, to accelerate it. There are many reasons for this and the architects themselves are only partially responsible since — as we shall see — they are by no means in full control of their own field of activity. Even so, their work shows certain characteristic deficiencies which they presumably could correct. One central reason for the drop in urban amenity is due to the fact that architects and urbanists alike persist in regarding the city as a largely visual phenomenon. Urban design is envisioned as being primarily a process of pictorial or plastic manipulation. As a result, the multidimensioned, psychosomatic reality of social life is violated by formalistic, two-dimensional designing. Another factor in this unhappy equation is that architects continue to design even enormous buildings as though they were to stand in a vacuum, assuming no responsibility for the ambiental repercussions of their designs. (This question of moral

Fig. 221. United States Embassy, Athens, Greece, 1961. Walter Gropius and The Architects Collaborative, architects.

Fig. 222. University of Baghdad, Iraq, begun 1960. Walter Gropius and The Architects Collaborative, architects. Working with a group, Gropius proved the continuing validity of his theories.

responsibility was at the heart of the famous controversy between the architects Walter Gropius and Pietro Belluschi and their critics over their role in the Pan Am Building in New York. But the country affords dozens of similar cases.)

The growing scope of the urban renewal program raises many architectural and urbanistic problems in acute form. This program was envisioned as being the means whereby life and amenity would be restored to the centers of our cities. How well is it succeeding? In many places — Boston, New York, Philadelphia, St. Louis, San Francisco — urban renewal programs are falling far short of their promised goals. And these failures derive in large part from the legislation upon which the entire program is built. For although it is heavily subsidized by federal, state, and municipal governments, the program is largely executed by private capital, operating under very inadequate architectural and urbanistic controls. Under such circumstances, major design decisions tend naturally to turn on the question of maximum profits. The projects are therefore escalated into the luxury class, with the social and economic composition of the "renewed" population quite different from the one it displaced. And yet, even when viewed within this escalated reference frame, most of these projects fall far short of their promise. They lack many of the authentic luxuries of urban life: clean air and clean grounds; first-rate public transport; freedom from dangerous traffic and ugly parking lots; a cosmopolitan choice of goods and services; streetscapes rich and interesting; parks and playgrounds safe and enticing in all seasons. For all their visual excitement, they are not truly luxurious.

Another facet of this same problem is to be observed in those new buildings in which monumental landscaped open spaces play an important role. Sometimes these are public projects — e.g., the new Boston state and municipal centers, the proposed FDR Memorial in Washington, the Federal Center in Chicago, or the Toronto City Hall; sometimes they are privately owned, like the plazas which surround the Chase Manhattan or Seagram skyscrapers in New York. In every case, stylobate or plaza, the architects have correctly envisaged them as features of the public world. How habitable, in real life, will these expensively developed open spaces prove to be?

Fig. 223. Edgar Kaufmann house, Palm Springs, Calif., 1946. Richard Neutra, architect.

Fig. 224. Warren D. Tremaine house, Santa Barbara, Calif., 1948. Though he has executed many projects, of all sorts, all over the world, Neutra's special talent is nowhere better exemplified than in the many splendid houses he has designed for various parts of the American Southwest.

The climates in which they stand are, without exception, difficult. Long, cold winters, full of wind, ice, and snow; intensely hot and humid summers full of heat, glare, and dust. What technical recognition is taken of these ambiental facts?

Unless provisions for snow removal or snow melting are built into these pavements, they will be covered with ice and snow for months on end — not the pretty snow of a Hiroshige lithograph but the soiled snow of modern cities — unsafe and unpretty, uncomfortably slushy, perilously icy. Contrariwise, how attractive will these same areas be in summer for the pedestrians to whom they are nominally dedicated? Are there trees, benches, fountains, restrooms, trash-baskets, adequate lighting — i.e., those amenities which make urban spaces truly urbane? These elements of the problem are not peripheral to the design: they are the essence of it. (Tacit admission of this is to be found in the fact that presentation drawings and official photographs are always made under optimum summer conditions.) Unless these new spaces are actually more attractive than the urban tissue which surrounds them, they will not draw and cannot hold the relaxed population of pedestrians which the architects envisioned.

Another factor which works against successful urban architecture is, ironically, our modern system of zoning ordinances. Conceived of half a century ago as a means of preventing urban deterioration, zoning theory assumed that diversity of land use was hostile to neighborhood stability. It therefore established a hierarchy of approved and forbidden types of tenancy and land use for each neighborhood, the theory being that homogeneity would halt decay. Time has amply proved the error of this theory (if anything, the rate of urban decay has *accelerated* since zoning became general). But the error lay not in the concept of control *per se* but in the criteria of control; these failed to make adequate distinctions between absolute nuisances (e.g., atmospheric pollution) and relative ones (e.g., residential units in business districts). Even more serious was the fact that middle-class ethical standards often replaced functional analysis. The city was classified according to "sacred and profane land uses" (to borrow Jane Jacobs' inspired phrase[8]), all the "better"

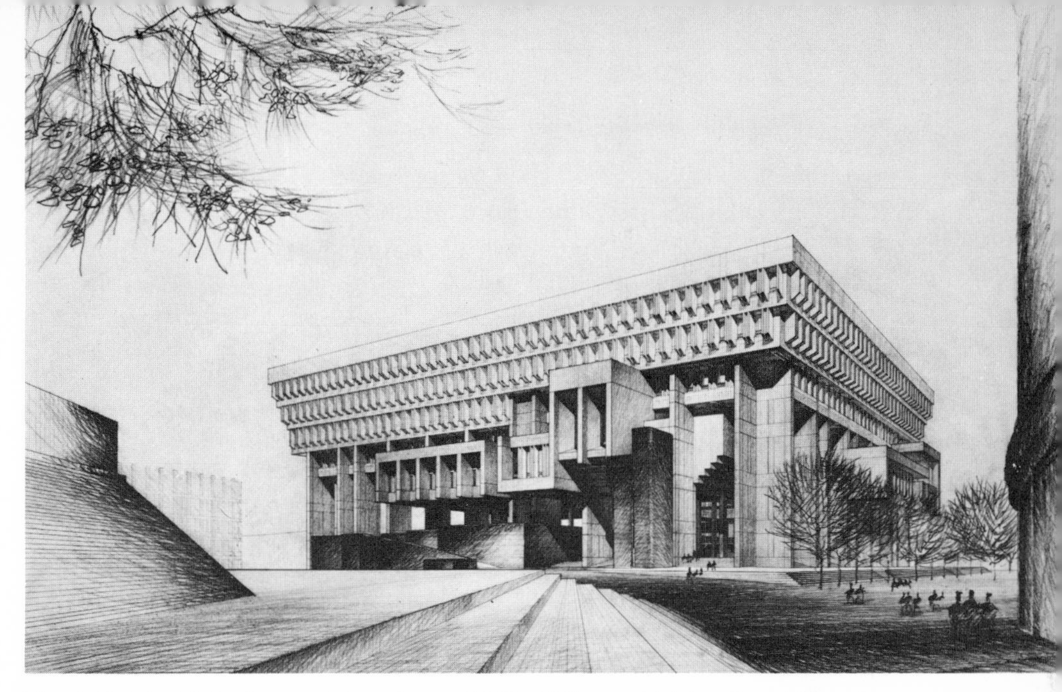

Fig. 225. City Hall, Boston, Mass., begun 1965. Kallman, McKinnell and Knowles, architects.

Fig. 226. Chase Manhattan Plaza, New York, N. Y., 1964. Skidmore, Owings and Merrill, architects. Isamo Noguchi, designer of the sunken fountain.

people being isolated from the less fortunate, the "good" stores from the less expensive, etc. The result of such policies was the creation of whole districts of monochromatic population and violent diurnal fluctuation. The upper-class American suburb is famous for its social and cultural imbalance and economic artificiality. Less well known but equally regrettable is the complementary imbalance of the central business district which supports it. Such areas are condemned to a disastrous half-life because zoning ordinances ordinarily permit a limited range of tenancies.

This policy of rigid segregation has negative side effects, too. For example, when it was built in the mid-fifties, the famous glass pavilion of the Manufacturer's Trust was a welcome addition to Manhattan's fashionable shopping district along Fifth Avenue. It was to exploit those shopping crowds that the new branch was located just there. It proved so successful that it started a rush of great banks into the area until, today, whole blocks of this part of town are occupied by pretentious banking rooms. But no amount of architectural virtuosity can offset the facts that banks generate little traffic of their own, are open only for limited periods each day and — open or closed — have very little interest for ordinary shoppers. Thus, too great a concentration of them tends to extinguish the very activity they seek to exploit. Inevitably, the crowds of shoppers which made the area desirable move elsewhere. A really wise zoning policy would recognize that there is an upper limit, a kind of saturation point, beyond which no single type of tenancy ought to be permitted to effloresce. A viable neighborhood requires diversity.

The central economic function of the city is that it affords man three essentials to civilized existence: proximity, predictability, and option. As I have pointed out elsewhere,[9] the physical expression of these is much more in the street than in the buildings which line it. Part of our persistent mismanagement of the city is due to our inability to understand this — and hence perceive the difference between the road and the street. The road is a means of moving people and goods from where they are to where they want to get to: but *a street is for people who are already where they want to be.* Thus the road can be almost indefinitely widened or extended. Since

Fig. 227. Fifth Avenue Branch, Manufacturer's Trust Co., New York, N. Y., 1954. Skidmore, Owings and Merrill, architects.

transport is its only function, it can be designed to accept any type of vehicle, in any quantity, moving at almost any rate of speed. But a city street, to be successful, must offer the proper ecological conditions for *homo economicus.* It is therefore by definition a pedestrian facility and must be scaled to his requirements in time and space.

Here, admittedly, we are on the borderline of present-day knowledge; we are dealing with supra-architectural qualities so little explored that they have as yet no accepted nomenclature, much less any accepted norms of critical evaluation. When urban renewal reaches a scale where whole sections of the city are reconstructed, we are no longer dealing with isolated architectural containers for one or another special function — housing, shops, schools — but with complex urban tissue in its entirety. Such tissue, to be viable, must support a whole spectrum of human need — social and private, somatic and psychic — which lies far below the reach of simple plastic or pictorial manipulation. And yet such superficial manipulation of urban forms can quickly inhibit, even drastically reduce, the life-supporting properties of urban tissue.

The fact is that urbane amenity is the end result of complex, interlocking processes, many of which are simply not the proper medium for artistic invention or subjective expression. These processes may not necessarily be "ugly" but neither do they "want" to be beautiful. We cannot expect urban tissue to grow spontaneously, by cellular division employing genetic memory, without human intervention; but neither can we expect such tissue to survive a formal manipulation for purely pictorial ends.

THE SKYSCRAPER IN THE FUTURE

The skyscraper is the building type which, more than any other, epitomizes modern American architecture to the world. And if we are to accept the continued expansion of the urban centers of the world as both inevitable and desirable, then it will become an increasingly familiar building type everywhere, for none other permits such a concentrated exploitation of land. But if American architects are to make a still wider use of this type in the future, then we should draw the proper conclusions from our experience with it in the past. And the first conclusion, obviously, is that few men have fully understood its implications. Louis Sullivan, as we have seen, recognized it as a new problem at the very start of its development; and, truth to tell, we have not carried it much beyond where he left it. But Sullivan's major concern was to give it, externally, a plastic expression congruent with its new structural and operational characteristics. He did anticipate the fact that when many skyscrapers were crowded together, new problems of a supra-architectonic scale would arise. But the building remained for him an isolated tower, self-contained and self-sufficient.

When does a building become a skyscraper? Man is the creature of gravity, his every action conditioned by its pull. His normal plane of intercourse is the surface of the earth and *any* discontinuity in that surface has profound effects upon his patterns of movement. Any

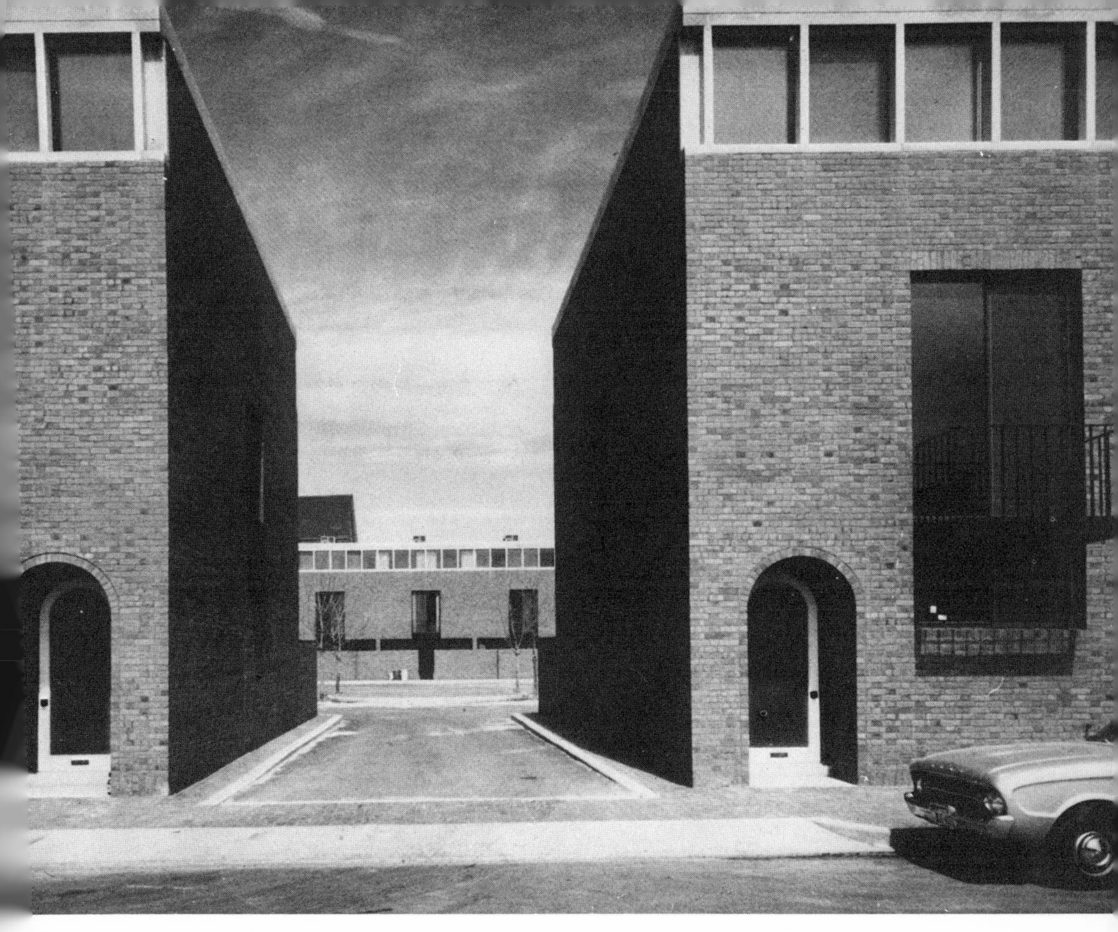

Fig. 228. Town houses, Society Hill Redevelopment Project, Philadelphia, Pa., 1962. I. M. Pei and Associates, architects and planners.

retailer knows that even a single step at the entrance to his store will significantly reduce the number of customers entering it; and the loss in trade will rise geometrically with the increase in steps. It requires a fairly strong motive to make a man climb a couple of flights. Five is about the upper limit of muscle and motivation. Above that, mechanical assistance is required. Above that point the building may be said to be a skyscraper.

Modern technology has made a commonplace of 40–, 60–, even 100–story towers. But it should not be assumed, as in the U.S. it has been assumed, that habitable space organized around a vertical plane of movement is qualitatively identical with space organized along a horizontal plane. Skyscraper life differs in many fundamental ways from life on the street. And these ways ought to be analyzed and understood much more precisely than they are today.

Irrespective of the type of tenancy (i.e., offices or apartments), the skyscraper has the effect of isolating the individual from the variety of external distractions which are normal to life at street level. Skyscraper life is life at the end of the cul-de-sac or the bottom of the well. The individual's freedom of movement and range of choice is greatly restricted. Even with high-speed self-operated elevators, a vertical trip of thirty floors down to street level is, subjectively, a much more significant act than a walk of three hundred feet down the street. One "thinks twice" before committing oneself to such a trip, even if the time and energy involved is less than a stroll down the block. Existence thus becomes compartmentalized, segregated. A high degree of concentration upon the task at hand becomes possible: from one point of view, this is productive but it introduces an unnatural "purity" into the day's experience. Life in the skyscraper is both objectively and subjectively different from life at the earth's surface, both for the husband at work in his office and the wife at her housework in the apartment.

The elevators themselves constitute a vulnerable link with the outside world (vulnerable in a sense the street could never be). This has been dramatically illustrated, during the last few summers in New York, when overloading of the electrical distribution system due to air-conditioning demands led to power failures. Thousands of people were trapped in the upper floors of office buildings, hotels and apartments. Hundreds of others were trapped in the elevator cabs themselves — often in express cars many floors away from the nearest doors — who had to be rescued by firemen with ladders and ropes!

Actually, these problems of the elevator are relatively simple: much more complex ones lie ahead. With the increasing use of automatic, self-service cabs, the elevator operator is disappearing and this has serious consequences in residential high-rise buildings. The unattended cabs become the scene of robbery, assault, and sexual molestation. In buildings without doormen or concierges (no public housing project ever had them) it is increasingly hazardous for women and children to use the cabs unaccompanied. Thus, this new vertical street, thanks to automation, has become as hazardous as

Fig. 229. Richards Medical Research Center, Philadelphia, Pa., 1964. Louis I. Kahn, architect.

Fig. 230. New Capital, Dacca, East Pakistan, begun 1962. Louis I. Kahn, architect.

Fig. 231. John McMullen house, Mantoloking, N. J., 1960. Marcel Breuer, architect; Herbert Beckhard, associate architect.

Fig. 232. Creative Arts Center, Colgate University, Hamilton, N. Y. (proposed), 1964. Paul Rudolph, architect.

the dark and unpoliced street of the Middle Ages! But even where these dangers do not exist, the tall building affords the family with young children a very unsatisfactory connection between home and the outside world. If the child remains in the apartment, he is arbitrarily isolated from his contemporaries. If the mother descends with him to ground-level playgrounds, she is separated from her adult tasks. In either case, the family is subjected to external stresses which would not occur in traditional patterns of low level housing.

Because elevators are expensive to install and to operate, there is a natural tendency to make them serve as many square feet of floor space as possible. In the office building, this may be unobjectionable. But in the residential tower, this deforms the plans of the individual units, robbing them of good ventilation, sunshine, privacy and view. It also leads to long interior corridors which are poorly lit, ill ventilated and noisy — often much less attractive than the streets which they replace. In this respect, even the most luxurious multistory apartment houses are seldom better designed than low-cost public housing.

Working conditions, no less than living conditions, are radically modified by the vertical organization of the skyscraper. The quality of life, at the level of both the district as a whole and the individual building, is determined by two phenomena — the great daily in-and-out movement of office workers into the district and the up-and-down traffic inside the skyscraper. These movements, occurring as flood tides four times a day, dump large populations into streets never designed to accommodate them. At rush hours, these are not large enough to contain the crowds, much less to afford all the facilities they need — cafés, druggists, shops, bars, etc. And provisions for relaxation, for sun and air, for any kind of light recreation such as softball, bocci, or checkers are nonexistent, as are drinking fountains, benches, and toilets. Difficult enough in fair weather, the situation of the average worker in bad weather becomes intolerable. There is literally "nowhere to go, nothing to do when you get there."

With its heavy wheeled traffic, narrow crowded sidewalks, solid walls, and open ends, the typical American street acts like a simple conduit. This form sets up a strong, linear current which is hostile to street life, both physiologically and psychologically. It creates a riptide along the face of the buildings at the very point where there should be quiet water, coves, and bays. The very nature of most human transactions requires the cul-de-sac, the enclave, the shaded portico and sunny courtyard — in short, the transitional zone between wholly open space and full enclosure. The current tendency to place the high-rise office building in a plaza of its own is commendable, so far as it goes. But it has not gone nearly far enough. Aside from being too abstract and generalized in design to be of much actual use to the office worker, even the best plaza does not resolve the problem of the street. This must be redesigned for the pedestrian, *for the man who is already where he wants to be.*

Clearly, the first step required to render the ground surface around the skyscraper habitable would be the banishment of all wheeled traffic, either to subterranean levels all its own or to the periphery of the district or — on a selective basis — private vehicles to the periphery and public transport belowground. But even without wheeled traffic the conventional street is totally inadequate to the needs of pedestrian traffic generated by clusters of office towers. What is called for here is a new kind of urban space, neither fully open like a street nor fully enclosed like a building. The actual design of such spaces would depend upon the local climate, first of all. In climates like those of Savannah, Las Vegas, or San Francisco, the problem would be technically simple. In fact, the loggias, arcades, and courtyards of historic architecture offer ideal prototypes against tropic rain or desert sun. In climates like those of New York, Detroit, or Duluth the problem of providing dry, wind-free, lighted, and partially heated spaces might be a little more complicated technically but even more urgent from the point of view of urban efficiency.

The dilemma posed by life in the skyscraper is essentially the separation it establishes between the public and the private aspects of existence. With the perfection of the modern elevator, this is no longer so much a matter of vertical distance above the earth *qua* earth as it is the separation from those public areas, services, and entertainments found only along the street. If this be so, then the skyscrapers of the future should be served (i.e., linked and pene-

Fig. 233. World Trade Center (project), New York, N. Y., 1965. Minoru Yamasaki, architect.

trated, like beads on a string) by aerial streets at every eight or ten stories. In office buildings, these aerial streets would offer the range of amenities and services required for a business district — restaurants, cafés, bars; specialty shops; areas for recreation and rest. In residential skyscrapers, the flying streets would have the normal facilities of the neighborhood — shops and services; playgrounds, crèches and kindergartens; movies, churches, etc. Public and private transportation — wheeled traffic of all sorts — would be carried by special streets at some intermediate plane, e.g., at the thirtieth floor for a group of forty-five story skyscrapers.

American experience with this building type furnishes another lesson for its future use: the policy of strictly segregated tenancies (commercial, residential, institutional, etc.) is mistaken. This has led to great high-rise housing projects that are little more than dormitory districts, filled and emptied twice a day. The skyscraper office districts have the same part-time use in reverse. The result is an enormously wasteful duplication of costly facilities. Each half of the community enjoys a kind of half-life, the entire system yielding far less in terms of true amenity than would a city of organically mixed uses. Thus, the upper half of future skyscraper cities might well be residential, the middle zone institutional, the lower zone commercial and manufacturing. In such an urban structure, daily commutation would involve a vertical trip of thirty stories instead of a horizontal trip of many miles.

The beginnings of such a new structuring of urban spaces as that outlined above can be glimpsed here and there around the world — in Helsinki, Cumbernauld, Coventry, Rochester. They were, of course, first visualized by Le Corbusier decades ago, with his famous lineal city proposals for São Paulo and Algiers. And they now engage the attention of architects and urbanists from Tokyo to Moscow.

THE CITY: JERUSALEM OR GOMORRAH?

Throughout the two decades following World War II, the erosion of the American city continued: the central city atrophied while "suburban sprawl" effloresced across the countryside. A contrary trend toward the reconstruction of the centers has only begun to make itself felt in the mid-sixties with the rising level of urban renewal programs. But the disastrous tide was by no means yet reversed; at best it was only slowed down. Nevertheless, increasing numbers of middle-class people began to question their flight to the suburbs. Many began to re-evaluate the city, even to return to it.

The situation was, to the highest degree, paradoxical. American culture in 1965 is the most completely urbanized the world has ever seen. More than 60 percent of its population lives in urban areas and — for the first time in human history more than 50 percent of its population belongs to the white-collar class (i.e., works in urbane occupations as opposed to manual labor in factory or farm). But it would not be possible to deduce this from our behavior or official attitudes. Far from acknowledging our absolute dependence upon the city, we behave as if its fate were inconsequential to our future. This misconception has been facilitated, ironically enough, by the urban revolution itself. Thanks to the spread of the internal combustion motor, the electric motor, and the electrical distribution grid the mechanization of American life in all its major aspects has been completed. Thereby the historic disparity between the material conditions of urban and rural life had been eliminated. A whole range of conveniences, amenities, and luxuries which had hitherto been available only in the city now became available in the countryside as well. Thus new modes of work and behavior have been introduced into the countryside which is thus converted into the theater of a much wider and more varied range of activities than would have been conceivable in preindustrial times. But this did not mean that the actual center of gravity of American life had shifted from urban to rural bases: quite the contrary, as is indicated by every index of American activity.

Historically, American attitudes toward the city have always been ambivalent. The migration of ambitious young Americans from the countryside to town has been continuous for a century or more. For them, no doubt, the city was the New Jerusalem, Horatio Alger's city of golden promise. But another important sector of American opinion, including many of our most important intellectuals and artists, has always viewed the city with mistrust: seen it as the center of vice, usury, and corrosive ugliness. From Audubon, Cooper, and Thoreau, there have always been important writers to celebrate the wholesomeness of country life while later muckrakers like Norris and Dreiser spelled out the perils of the city. American folk and genre painters extolled the beauties of bucolic farm life or noble frontier; and not until the end of the nineteenth century did the urban scene begin to engage the attention of our major artists. Even the architects and landscape architects, for all their urban origins and urbane occupations, sang the beauties of the rural (or at least suburban) scene. A line of romantic protest against city life runs from Jefferson through A. J. Downing to Frank Lloyd Wright.[10]

Fig. 234. Garden, Museum of Modern Art, New York, N. Y., remodelled 1964. Philip Johnson, architect.

Fig. 235. Redevelopment Scheme, Fort Worth, Texas, 1958. Victor Gruen Associates, architects.

The inherent irony of these attitudes lies, of course, in the fact that the artists and intelligentsia are by definition urban phenomena.

Nevertheless, loved or hated, the importance of the city was at least *recognized* until the end of World War II. But increasingly since then, American attitudes have undergone a profound change. More and more the city came to be described as "doomed," "obsolete," "disappearing." The middle and upper classes fled to the suburbs: while maintaining their economic ties with downtown, they relinquished all civic interest and responsibility. The city, historically the residence of the rich and mighty, had become the province of the poor. And the poor, not owning the city and hence having no real control over its destiny, retreated into privacy. As one observer has put it:

> . . . the New Yorker's experience of "belonging" consists largely of jostling his neighbor in the barbaric squalor of the subway. He naturally retreats into privacy because his public encounters — on dirty streets, in mean civic buildings, in truncated railroad terminals, on overcrowded buses, in under-policed parks — are so deliberately shabby. His withdrawal is so total that he can tolerate the smell of urine in the subway corridors and ignore the obscenities scrawled in his apartment house lobby. Thus he denies cognizance of what he does not own, rather than create with his ample resources a public environment that is, in the root sense, *civilized*.[11]

Parallel to the steady attrition of public amenity in the city has been the efflorescence of the middle- and upper-income suburbs and the privatization of material well-being which they represented. In contrast to the increasing ugliness, confusion and squalor of downtown, these suburbs afforded a physical environment of beauty, privacy, safety, and material comfort. These advantages were so overwhelming that, for decades, they served to conceal a range of built-in deficiencies. Only a generation's experience has begun to reveal them. Though they derive from the actual physical structure of these suburbs, these deficiencies are expressed in psychological, social, and ultimately cultural terms. The most serious limitation, from a social point of view, is segregation: this always operates economically even when not present explicitly in ethnic or religious terms. The cultural result is a parochial society — one-dimensional, mono-colored, politically alienated; in a word, parasitical.

For the individual member of the suburban family, the usual consequences are restricted motility, a narrowed range of option, isolation from the society of his peers. This phenomenon is a function of the suburb's large-scale layout, thinly spread population, and highly restricted land uses. Unlike even the more densely knit texture of the earlier streetcar suburb, the automobile suburb inhibits movement on foot, by bicycle or public transportation. Every movement depends upon the private automobile. Yet only a portion of the suburban population can actually enjoy the putative independence it affords: young children, adolescents, the aged, the handicapped and the sick are dependent upon the able-bodied adults of the family. With teen-age children away at school and the husband away at work, the housewife becomes the chauffeur for the family: what mobility they thereby win is thus at the cost of her freedom.

But perhaps the most disturbing aspect of suburban life is the artificially compartmentalized existence it imposes upon the family members. It projects the housewife into a narrow and repetitive orbit of tasks and responsibilities which are partly servile and partly stultifying in their endless repetitiveness. It locates the children in a Potemkin village of conformity, restricting their contacts to children of their own ethnic, economic and cultural milieu and, at the same time, isolating them from any contact wtih productive adult activity. Finally, the suburb catapults the man into a schizoid existence, migrating twice daily between a public and a private world, each with its mutually contradictory values. It deprives his children of the possibility of understanding their father's function in the real world. This migratory existence weakens his civic allegiance to either suburb or city.

Obviously, the American city, street, skyscraper and suburb all fell far short of satisfactory performance in 1965. And fundamental to their malfunction was the still unresolved problem of transportation. In no other area of life was the drop in public amenity so marked. A great system of steam trains, electric interurbans, trams, elevated and subway systems had been created in the century 1830–1930, as had a complementary system of river, lake, and coastal steamers. This had given the U.S. the finest public transportation in the world.

But in the last three decades this system has been allowed to fall into decay until today mass transport — especially the passenger-carrying facilities — is worse in many parts of the nation than it was a century ago! (Not a single passenger train remains in the state of Maine — an area larger than Ireland. It is impossible to cross Los Angeles in one uninterrupted trip except by private auto. There are great cities in Texas with no public transport except a skeletal bus system largely designed to carry Negro domestic servants from the slums to the outlying suburbs.)

All this has been replaced by an inconceivably fragile, inefficient, and expensive network of highways and airlines. (While the Congress jousts with windmills about aid to failing railroads, the federal government has spent some forty billion dollars on a highway program and many billions more on airport facilities.) The imbalance and fragility of the new system is frightening. (A recent fog in the New York metropolitan area led to the cancellation of 1400 out of 1780 flights. Planes from Paris had to land in Detroit. One plane from Dallas made two round trips to New York before finally landing in Washington.[12]) And the cost is incalculable, both publicly and privately. (Over two-thirds of the land area of downtown Los Angeles is devoted to facilities for the private automobile — streets, garages, parking lots.[13] One can fly from New York to Palm Beach in less time than is required to reach the airports of the two cities.)

New architectural facilities for automotive and plane traffic are often visually spectacular. But designs like Nervi's for the Port Authority Bus Terminal in Manhattan or Saarinen's terminals at Kennedy and Dulles Airports give a quite misleading image of the modernity of the system as a whole. (Despite the large number of passengers using them, only one airport in the nation — Boston — is connected to the central city with rapid mass transport.) The auto and plane are essential parts of modern life: they only become grotesque when cultivated at the expense of a rationally designed system covering all aspects of intra- and inter-city transportation. Instead of simplistic commitment to car and plane alone, we ought to be employing the whole spectrum of technically advanced devices: jet-powered monorail, linear induction motor, hydrofoil lake and river boats, and moving sidewalk and escalator.

Fig. 236. TWA Terminal Building, New York, N. Y., 1961. Eero Saarinen and Associates, architects.

Fig. 237. Terminal Building, Dulles Airport, Washington, D. C., Eero Saarinen and Associates, architects.

All of the factors described above converge to render American architecture less effective *in toto* than its individual components seem to promise; and incomparably less urbane and less beautiful than we have the right to expect of a technology like our own. In our anxiety to reach the other planets, we are sadly neglecting this one. This neglect is expressed in polluted urban atmospheres and waterways (in 1964, New York City was still dumping half a billion gallons of sewage per day into the waters around it; and Miami Beach still dumped all its raw sewage into the very waters on which its whole economy was based!); in despoiled countrysides; in wasteful traffic jams and decrepit public transport. And this all-too-visible confusion and squalor is paralleled by invisible deficiencies in classrooms, hospital beds, housing units, and recreational facilities. It is not only that we do not reach our self-proclaimed goals: it is even more that, given American potentials, the goals themselves are far too low. Under such circumstances, the American architects have special obligations to their society. How well prepared are they, as an organized profession, to discharge them?

THE PROFESSION: ITS STRUCTURE AND FUNCTIONS

At the end of 1964, the profession consisted of 32,814 registered architects.[14] In addition to one national association, the American Institute of Architects with twenty-nine state chapters, there were eighteen state societies all of whose members were registered architects but not all of whom were members of the AIA. In fact, though the AIA was the unchallenged spokesman for the whole profession, its membership stood at only 17,516.[15] There were, in addition, about 3500 graduate professional apprentices — i.e., young men and women who have won their first professional degree but not yet completed the three-year stint of supervised work in licensed offices.[16] This apprenticeship is the equivalent of the medical internship and, like it, the precondition to standing for the state board examinations.

Admission to the profession is won, as in medicine or law, by passing these examinations, held by the authorized agencies, the Boards of Architectural Registration of the various states. Operating under state laws deriving their powers ultimately from the public health and welfare clause of the United States Constitution, these quasi-governmental bodies (the first of which was created by the Illinois Legislature of 1897) exercise effective control over the entire profession.

In 1964, the profession supported a total of seventy-seven schools or departments in as many colleges or universities: these were organized nationally into the Association of Collegiate Schools of Architecture.[17] The academic standards and performance of these schools was, in turn, policed by another professional organ, the National Architectural Accrediting Board. It had fully accredited fifty-three of the above schools.[18] It reported a total enrollment of 18,922 students during the academic year 1963–64.[19] Of these, the great majority would have been working for their first professional degree. Figures on higher degrees were not available but the number is steadily increasing, as in all professions.

Structurally, in other words, the architectural profession closely resembles those of law and medicine. There are several significant differences, however. First, the architect is not in such complete control of his field as are his peers in medicine or law. The civil engineer is also qualified, by the terms of his education and licensing, to design structures involving public health and safety. Most bridge and highway construction, as well as airports, dams and seaport facilities, is ordinarily entrusted to the civil engineer; the architect will be involved (if at all) only in the monumental aspects of the projects. In addition, whole sectors of specialized industrial and military enterprise are considered the exclusive domain of the engineer.

Other sectors of the building industry lie outside the architect's normal jurisdiction. Many aspects of urban planning are now entrusted to professionals with nonarchitectural backgrounds. Then there is the twilight zone of the "stock plan" in which the architect designs a prototype — chain store, filling station, single-family house — for serial production. Here the client will be a national chain or a big speculative builder; and the architect will be paid on a royalty basis, having little to do with the actual siting or construction of the individual building. Most of the huge single-family housing

Fig. 238. New Office Building (project), A.I.A. Headquarters, Washington, D. C., 1965. Mitchell and Giurgola, architects. This design placed first in a national competition run by the A.I.A.

market, amounting to around a million units a year, is handled in this fashion. Finally, there is the fact that, in most rural areas of the nation, any individual can build his own house or barn without recourse to either architect or engineer since the health and safety of the public is not involved.

This lack of control over his putative field of operations unquestionably makes the architect less influential than the medical doctor, whose monopoly of power often seems absolute even in such peripheral fields as pharmaceuticals and health insurance. Because of such factors, the ratio of architects to engineers or lawyers is much lower per capita (26.5 per 100,000 as against 449.8 and 203.8 respectively). The median income is also lower ($5509 for architects as opposed to $8302 for doctors and $6284 for lawyers).[20]

But the architectural profession has other distinguishing characteristics as well. The traditional firsthand relationship between individual practitioner and client is increasingly being modified as both tend to become corporate entities. Even as early as 1950, 24 percent of all registered architects were salaried employees, either of big private firms or of governmental agencies.[21] The trend is increasingly pronounced: ever fewer architects stand any real chance of becoming successful free-lance practitioners. Abstractly, this may not of itself be a bad development, but it has two important implications for the future. First, it means that the employed architect in the big office will become increasingly specialized, seeing less and less of the entire field of practice and of the clients themselves. Second, it implies that more and more of his clients are themselves corporate entities — big business, big institutions, big government.

The implications here are disturbing, in that the architect's *legal* client is less and less his *real* client. The white-collar worker in the big skyscraper, the blue collar worker in the big plant, the housewife in the big housing project, the child in the big consolidated school — these are the people for whom the architect works and to whom he is ultimately responsible. His buildings should be accurately and sensitively designed for them. Yet these are the people whom the architect today seldom sees. He deals instead with their agents — those corporate or institutional entities who are the legal owners of the projects. Instead of firsthand observation of real people, intimate contact with their needs and aspirations, the architect is given statistical data about them — peak loads, median incomes, average family size, minimum floor areas, etc. These data may be essential for the establishment of the broad lines of policy; but they are no more a substitute for firsthand knowledge of the actual inhabitants of the building than statistics on the incidence of cancer would be to the physician with an actual patient.

It cannot be assumed that these corporate or institutional clients, acting though they may be as legal agents for the consuming public, are always to be relied upon to represent its best interests or optimal requirements. In a profit-motivated society, criteria for architectural projects are all too apt to approach the minima permitted by building codes or the law of supply and demand. Of course, this tendency will vary from field to field. The speculative builders of tract housing are notorious for the venality of their design and construction policies. But a life insurance company, planning a big housing project for long-term investment purposes may — out of sheer self-interest

— follow quite enlightened policies. Moreover, the demand for maximum economy in public projects can often prove as limiting a factor as the demand for maximum profits in a private one. Many public buildings with specialized requirements, such as hospitals or schools, must meet objective standards determined by the uses themselves. American public housing, on the other hand, has been permitted to slip to the lowest levels of mediocrity of any country in the West.

The building consumer is thus unable effectively to express his demands, requirements, expectations; and the architect's isolation from them leads inevitably to the abstract, the formal and the platitudinous in architectural and urban design. His changed relationship is not only expressed in the architect's personal life, whose fortunate position in Galbraith's economy of abundance has served to insulate him against the squalor and deprivation of Harrington's culture of poverty.[22] It is also expressed in changes in his cultural orientation. By the very nature of his work, the architect has always stood closer to the rich than to the common man, for whom his services were always less imperatively required than those of the doctor or lawyer. Nevertheless, the tradition of the socially conscious, intellectually committed architect has had a long history in the United States. Indeed the leading spokesmen for the profession in each generation tended to that persuasion: Jefferson, Bulfinch, Latrobe, Greenough; Olmsted, Sullivan, Wright; Clarence Stein, Gropius, Neutra. This tradition saw its apex during the days of the Great Depression and the New Deal, when unemployment in normal channels forced approximately four-fifths of the architectural and engineering professions into government-subsidized projects of one sort or another. This switch from the private to the social client gave the architect an exhilarating sense of identity with society. It served to shift his attention, if not his allegiance, toward social architecture.

The afterglow of the Rooseveltian liberalism still suffuses the profession, giving a more liberal aspect to its posture than, perhaps, those of the American Medical Association or the American Bar Association. But increasingly since World War II, the profession has returned to "normal." Architects have been reabsorbed into the world of private enterprise. Some of them have themselves become men of big business: in 1964 there were twenty-five firms *each of which* had over $60,000,000 worth of work on its boards.[23] Ironically, this general prosperity has led to the impoverishment of intellectual speculation and artistic invention. The utopian element in architectural thinking has been largely submerged. Theory itself is in disrepute. The men whose polemics (whether in print or in stone) once galvanized the Western world are either aging or gone. And though Mies van der Rohe and Gropius are still active and still vocal, few younger men aspire to their prophetic role. The dominant attitude is one of complacent *laissez-faire,* the esthetic expression of which is a genial eclecticism.

The result of all this is a body of work whose general orientation is as antipopular or aristocratic as that commissioned by Frederick the Great or Louis XV. This aristocratic esthetic is very evident in the most prestigious buildings of the day. It is dismayingly apparent in urban redevelopment work, where almost without exception the poorer worker, the racial minority, the small merchant and businessman is replaced by "upper middle-income" groups. These projects are often handsome from a purely formal point of view. But in substance they represent the sort of "class planning" employed by Baron Haussmann when he remade central Paris to correspond to the imperial pretensions of Napoleon III. The novelist Norman Mailer was quite correct when he charged that, even as a mode of stylistic expression, this architecture had been emptied of its functional-democratic-progressive contents.[24]

The spacious and humane iconography of the New Deal is being put to quite other uses today; but it will not be the first time that the same style has been the vehicle of different points of view. As we have seen (pp. 62–72), Jefferson and Napoleon alike turned to the Romans for a language of figurative expression; and two such antithetic figures as Emerson and Calhoun could both feel comfortable with the Greek Revival. Nevertheless, whatever one's personal estimate of this aristocratic esthetic, it is still the outward evidence of an internal involution: the abdication by the architect of his claim to speak for the whole people, to become instead the ideologue for an elite.

How the profession will extricate itself from this cul-de-sac remains to be seen. Because of default, it probably faces that sort of "socialization" the fear of which has convulsed the medical profession, and for many of the same reasons. For, just as millions of Americans lack good medical care, so the same millions are deprived of good hospitals, schools, and housing. The satisfaction of this backlog of need implies increased intervention of governmental agencies. This, in turn, probably implies an increase in the bureaucratic architecture so abhorred by the profession. But — aside from the fact that big private firms are quite as bureaucratically organized as any counterpart in government — this "socialization" of architecture will not, of itself, guarantee qualitatively superior architecture or town planning. This depends on many factors, not least of which would be a greatly improved system of education, for building consumer and building designer alike.

A NEW SORT OF EDUCATION

The historical origins of the architect's training lie much closer to those of the engineer or dentist (both of whom began as "mechaniks") than to those of doctor or lawyer. This was due to the fact that competence in the former fields *could* be acquired through apprenticeship alone, whereas theoretical training was always considered essential in the latter. Thus, until the opening of the present century, the building field was largely in the hands of men whose origins were closer to the craftsmanship of millwright and mason than to academic scholarship. Professional education at the university level has been a commonplace in medicine and law for centuries. But the first American school of architecture was established only in 1866. (The AIA itself had been formed only nine years earlier in an effort to establish professional standards of training and competence.) It was to be decades before the last state government could be persuaded to establish a system of examination and licensing procedures (Illinois was first in 1897, Vermont last in 1951).[25]

But preindustrial architecture had one anomaly which it shared with no other profession — the presence in its midst of the amateur and the connoisseur. (Both terms, in those days, had the very favorable connotation of a disinterested love of the field.) From the Renaissance onward, the architect's patron often crossed the line to become, in actual fact, an architect himself. But whether these amateurs remained passionate and tireless patrons, like Horace Walpole or the Earl of Burlington, or became the actual designers of buildings, like Jefferson at Charlottesville, they established the other polarity of preindustrial architecture. As opposed to the direct economic incentives of the craftsman, they entered the arena from many other approaches: from an enthusiasm for literature, especially the classics; from antiquarianism; from political commitments (Jefferson saw the Roman basilica as the essential container for his new Republic); even from religious fervor (the Gothic Revival was above all the vehicle for nineteenth-century religiosity).

Formed thus of components from widely divergent cultural milieus and propelled by the most disparate motivations, it must be reckoned as extraordinary the extent to which these early architects managed to resolve the contradictions between craftsmanship and scholarship, poesy and practicality — in short, the contradictions between the formal and the functional. Nevertheless, the fundamental tensions remained: in fact, they grew steadily sharper with the rise of industrialism; and they are accurately reflected in the curricula of the schools today. Generally speaking, the response of the schools has been to place increased emphasis on the academic, always at the expense of the craft component of the field. From one point of view, this has been at once inevitable and desirable. Modern architectural problems can no more be solved by carpentry than could space craft be built by blacksmiths. The shift in emphasis away from craftsmanship, however, has been more toward technology than toward a truly scientific investigation of architecture.

There can be no denying that the price paid for this new professionalism has been high. While it has made possible a new order of building performance, it has simultaneously permitted the categories of error discussed earlier in this chapter. It has meant the demise of the rich conventional wisdom of preindustrial times. Few contemporary architects have any firsthand knowledge of actual construction methods and techniques. And the same process robs the craftsman of his historic competence and wisdom. What with the factory production of building components and the mechanization of the construction industry itself, there is less and less need for the intelligent, well-trained craftsman. Indeed, it has been argued that modern working drawings from the architect's office tend objectively to *discourage* good workmanship. Increasingly, as they reduce building to a process of mere assembly, they imply the "headless hand" of the assembly line. The result is the further lowering of the motivation, taste, and literacy of the workman, a situation which the old apprenticeship training system made impossible. And this disastrous circle of impoverishment is closed by the disappearance of the individual client and his replacement by the faceless mass consumer of architecture, who has no real voice in the design or control over the operation of the buildings in which he is born, lives, works and dies.

The new education cannot, alone, reverse this trend but it can play a decisive role. Moreover, it is the area over which the profession exercises complete control and in which it can act decisively. And the first thing this new education should do is to define, in terms much more precise than ever before, the nature of architecture. For, unlike most other fields of professional activity, architecture serves a complete spectrum of social process which stretches from the ceremonial (e.g., the tomb of President Kennedy) to the wholly utilitarian (e.g., the surgical operating theater). Because its jurisdiction thus extends from the poetic to the practical, it is simultaneously subject to two quite different value systems — one derived from art, the other from science and technology. This confronts the architect with problems which are unique to him.

Subjectively, the architect occupies a position closer to the artist than to the scientist in that he aspires to the creation of *formal* order. That is, like all artists, he seeks to resolve the contradictions between form and content in such a way as to extract from it a work of high esthetic value. Here, however, the parallel ends: for the architect hopes to achieve a building which will have an artistic value *above and beyond the basic need which called it into being.*

This dual ambition makes his task, if not of a higher order than that of his fellow artists, then certainly one of far greater complexity. For though his building may, like a piece of sculpture, be susceptible of manipulation for purely formal ends, the content of his work is wholly different: social process and live human beings, *each with ineluctably nonesthetic requirements.* This difference can be formulated in another way. The relationship between the painter's canvas and its audience may be described as a dialogue — visually dimensioned only, limited in scope and commitment on both sides, easily terminated by either party. But the relationship established between the architect's building and its occupants is of quite another order. It is environmental, uterine in both extent and profundity. No modern man can wholly escape it and it is terminable only by death.

Thus, whatever formal characteristics a work of architecture may share with other categories of art, there remains this fundamental difference: *architecture has no spectators, only participants.* (The weakness of most architectural criticism is precisely that it is based upon pictorial *representations* of the building, not upon experiential immersion in the building itself. This isolation of the visual dimension from the multiform matrix of reality deforms our critical literature and seriously reduces its value.) This distinction, so obvious and yet so profound, is too often ignored by everyone, including the architects themselves. All too often, as a result, any contradictions between the formal requirements of the container and the functional requirements of the contained is arbitrarily resolved in favor of the former. The occupant is simply forced to fit.

As though this were not enough to differentiate the teaching and practice of architecture from all other forms of artistic activity, there is another paradox concealed within the "contained" itself. The environmental requirements of the occupant of the building often differ fundamentally from the various processes which, in the modern world, he is expected to attend, take part in, supervise. In the preindustrial world this inner contradiction was not usually severe but in modern technology these two sets of environmental requirements are not always congruent. In modern industry, they are widely divergent: in such industries as printing, pharmaceuticals, or metallurgy, environmental criteria for worker and process will differ radically. Often — as in chemicals or nuclear power — they will be mortally irreconcilable. Thus, in many of the 270 distinct building types which contemporary American life requires,[26] contradictions must be resolved at two distinct levels: first between the persons and processes contained and then between them and their container. Respect for these two conditions is mandatory if the building is to be operationally successful. And yet respect for these two conditions by the architect will often leave very little room in which he can manipulate the container for purely formal ends.

Most contemporary failures in architecture and urbanism stem either from a failure to understand this situation or else from a refusal to come to terms with it. Of course, as we have said, neither buildings nor cities can grow like living organisms. They require the intervention of human agencies. In this sense, they must always be consciously designed by men and these men will always have preconceived ideas of what forms they should assume. An "unplanned" edifice or city is ultimately inconceivable: no matter how badly designed or executed, it is always the expression of somebody's creative ambitions. But, more than ever before in history, these ambitions must be contained, disciplined and structured by objectively verifiable terms of reference.

Such problems have always been present in architecture, of course; but for us they have been acutely complicated by the flowering of science and technology, on the one hand, and the unparalleled proliferation of social, economic and industrial processes on the other. It should be remembered that, at any time prior to the closing years of the eighteenth century, the architect would have worked within a clear and comprehensible reference frame of needs and means. The exact shape and dimensions of this frame were established by two sets of complementary but opposing forces. One set of forces pressed outward, so to say, as the expression of the client's minimal requirements. These minima would vary with class and culture. Thus they would extend from the peasant's irreducible requirements for survival — roof against sun or rain, wall against heat or cold, enclosure for crop and beast — to the satisfaction of the sensual appetites and ceremonial apparatus of the aristocracy. But another, external set of forces acted to limit the demands for survival and satiety alike. These pressed relentlessly inward, established limits of size and scale, span and height, complexity and technical sophistication. They sharply restricted the building activities of pharaoh, emperor and pope alike.

The external, inward pressing forces were six in number:

(1) the impact of the climate
(2) a limited palette of locally available building materials
(3) the lack of mechanical prime movers
(4) the limited means of transport and communication
(5) a slow rate of cultural change and invention
(6) an extremely well-informed but conservative clientele

Under such restrictive circumstances, factors of safety and margins for error were small; personal idiosyncrasy was sharply curtailed; the efficiency of the artifact easily established by society; the range of possible solutions to any given problem limited by existing resources. Under such circumstances, an architect like Thomas Jefferson could readily encompass the poetic and practical aspirations of his society, designing both the curriculum and the campus of his beloved university and then supervising the making of bricks and nails to construct it.

In 1965 this balance of forces has been radically altered.[27] Industrialization is making possible human habitation in any environmental circumstances, not excluding the ocean bottoms and outer space. The natural materials of the whole world and a range of synthetic new ones are available to the architects of all developed countries. The availability of mechanical prime movers for all tasks is, for all practical purposes, limitless. Communication at the speed of light and transport faster than sound have raised the rate of cultural change to unprecedented levels. The result is a clientele open to radical change, insatiable in its appetites but quite illiterate in its judgments.

It is apparent that the external limitations which held historic architecture, so to say, "in shape" are no longer operative. They have disappeared forever and must be replaced by a new internally generated discipline. This discipline can only derive from the most scrupulous re-examination of man and his relation to his environment; and this examination must fully utilize both the norms and the substance of modern scientific knowledge. It will be the task of the second volume of this work to indicate the direction and scope of that examination.

Fig. 239. Chapel of the Holy Cross, Sedona, Ariz., 1953. S. Robert Anshen (1910–1964) and William S. Allen, architects.

NOTES

CREDITS FOR ILLUSTRATIONS

INDEX

NOTES

CHAPTER 1
1620–1776: WHAT WE HAD TO BEGIN WITH (pages 1–36)

1. Edward Johnson, *History of New England* (1653). Cited by M. S. Briggs, *Homes of the Pilgrim Fathers* (New York: Oxford University Press, 1933), p. 122.
2. John Smith, *Advertisements for the Unexperienced Planters of New England, or Anywhere* (London, 1631).
3. *A Description of New England: or the Observations and Discourses of Captain John Smith* (London, 1616).
4. The exact development of the nail, from its handmade origins to its modern mass-produced form, has not yet been thoroughly investigated. The most authoritative article to date is "Nail Chronology as an Aid to Dating Old Buildings," *Technical Leaflet No. 15,* U.S. National Park Service (Philadelphia: 1963).
5. *Records of the Colony of Massachusetts in Massachusetts.*
6. Letter to Dr. Ingenhousz, written at sea, August 28, 1785.
7. John Shute's *First and Chief Groundes of Architecture* (1563) was the first English attempt to bring the Classic Orders to English builders.

CHAPTER 2.
1776–1820: THE NEW REPUBLIC RISES (pages 37–73)

1. *Letter,* Washington to Johnson, Congressional Documents, 19th Congress, Report No. 228, p. 28.
2. Brooks Adams, Introduction, *The Degradation of Democratic Dogma* (New York: Capricorn Books, 1958), p. 19.
3. Turpin C. Bannister, "Architectural Development of the Northeastern United States," *Architectural Record* (June 1941), p. 68.
4. In his book *Sketches Historical and Descriptive of Louisiana* (Philadelphia: Mathew Carey, 1812) Amos Stoddard writes (p. 161) that "forty-two miles above Plaquemine is the first saw mill . . . No other wood than white and yellow pine [is] sawed, and the last [is] deemed the best for most purposes."

5. Letter to Trustees of East Tennessee College, May 6, 1810.
6. *Notes on the State of Virginia,* 1782.
7. *Ibid.*
8. Letter to Dr. Thornton.
9. *Objects for Attention for an American,* June 3, 1788.
10. Letter to Col. Humphreys, August 14, 1787. *Works* (Princeton 1955), vol. 12, p. 32.
11. Thomas Paine, *The Age of Reason:* Part I (New York: Vincent Parke and Co., 1908), vol. I, p. 64.
12. Milton Brown, *Painting of the French Revolution* (New York: The Critics' Group, 1938) p. 80.
13. *Journal of Latrobe* (New York: D. Appleton & Company, 1905), p. xv.
14. *Ibid.*, p. 139.

CHAPTER 3

1820–1840: QUIET BEFORE THE STORM (pages 74–106)

1. Henrietta Buckmaster, *Let My People Go* (New York: Harper and Brothers, 1941), p. 251.
2. Charles Dickens, *American Notes* (Greenwich, Conn.: Fawcett Publications, Inc., 1961), p. 264.
3. Cited by James Marston Fitch, *Architecture and The Esthetics of Plenty* (New York: Columbia University Press, 1961), p. 66.
4. Winton Rowan Helper, *The Impending Crisis of the South: How to Meet It* (New York, 1860), p. 21.
5. Van Wyck Brooks, *The Flowering of New England* (New York: E. P. Dutton and Company, 1936), p. 183.
6. The most comprehensive account of the architecture of the period is to be found in Talbot Hamlin's *Greek Revival Architecture in America* (New York: Oxford University Press, 1944).
7. J. Hector St. John de Crèvecoeur, *Letters from an American Farmer* (New York: Fox, Duffield, 1904), quoted by David R. Weimar, *City and Country in America* (New York: Appleton-Century-Crofts, 1962) pp. 8–9.
8. *Ibid.*
9. John Mason Peck, *Forty Years of Pioneer Life,* Rufus Babcock, Editor (American Baptist Publication Society, 1864), p. 144.
10. John Mason Peck, *A New Guide for Emigrants to the West* (Boston: Gould, Kendall and Lincoln, 1837), pp. 122–24.
11. *Ibid.*, p. 125.
12. Vernon Parrington, *Main Currents of American Thought* (New York: Harcourt Brace and Company, 1930), vol. II, p. 99.
13. *Ibid.*, p. 102.
14. According to Talbot Hamlin, "Greek Revival and Some of its Critics," *The Art Bulletin* (September 1942), p. 252.
15. *Journal* of the Franklin Institute (January 1841), vol. XXXII, no. 1, p. 17.
16. H. M. P. Gallagher, *Robert Mills* (New York: Columbia University Press, 1935), p. 156.
17. Philip S. Foner, *History of the Labor Movement in the United States* (New York: International Publishers, 1947), p. 171.
18. Charles A. Madison, *Critics and Crusaders* (New York: Henry Holt and Company, 1947), p. 123.

19. Robert Dale Owens, *Hints on Public Architecture* (New York: G. P. Putnam's Sons, 1849), p. 48 ff.
20. George H. Weitzman, in his paper "The Utilitarians and the Houses of Parliament" (*Journal* of the Society of Architectural Historians, October 1961, vol. XX, no. 3, pp. 99–107) has given a clear account of the way in which such questions were handled by the committees in charge of the work.
21. Henry Cleveland writing in *The North American Review* (1836). Cited by John Burchard and Albert Bush-Brown, *The Architecture of America* (Boston: Little, Brown and Company, 1961), p. 97.
22. George Templeton Strong, *Diary,* edited by Allan Nevins and Milton H. Thomas (New York: Macmillan, 1952), vol. I, p. 240; vol. I, p. 234; vol. II, p. 328; vol. I, p. 261.
23. Dickens, *American Notes,* p. 228.
24. See Fitch, *Architecture and the Esthetics of Plenty,* pp. 173–188, for a more extensive analysis of Western attitudes toward nature.

CHAPTER 4

1840–1860: THE SCHISM (pages 107–139)

1. See Fitch, *Architecture and The Esthetics of Plenty* (New York: Columbia University, Press, 1961), pp. 65–85.
2. *Ibid.*
3. Richard O. Cummings, *The American Ice Harvests* (Berkeley: University of California Press, 1949).
4. James Fenimore Cooper, *Correspondence,* edited by his grandson (New Haven: Yale University Press, 1922), p. 613.
5. Cited by Robert E. Riegel, *Young America,* (Norman: University of Oklahoma Press, 1949), p. 43.
6. George Templeton Strong, *Diary,* edited by Allan Nevins and Milton H. Thomas (New York: Macmillan, 1952), vol. II, p. 399.
7. R. C. Carpenter, *Art of Heating and Ventilating Fifty Years Earlier* (New York, American Society of Heating and Ventilating Engineers, 1905).
8. Catharine E. Beecher and Harriet Beecher Stowe, *The American Woman's Home,* (New York: J. B. Ford & Co., 1869), p. 24.
9. Walker Field, "Origins of the Balloon Frame," *Journal of the Society of Architectural Historians* (Amherst, Mass., October 1946), pp. 5 ff.
10. The American Society of Civil Engineers, oldest national engineering society in the United States, was founded in 1852.
11. Two recent studies help to explain some of the paradoxes of this tormented genius: *John Ruskin* by Joan Evans (Oxford University Press: 1954) and *The Darkening Glass* by John D. Rosenberg (New York: Columbia University Press: 1961).
12. John Ruskin, *Collected Works,* (New York: John W. Lovell, 1885), vol. III, p. 16.
13. *Ibid.*, p. 18.
14. Ruskin, *Works,* vol. XVIII, p. 502.
15. *Ibid.*, p. 443.
16. See Fitch, *Architecture and the Esthetics of Plenty,* pp. 46–64, for an analysis of Greenough's philosophy of art. A selection of his essays, edited by Erle Loran, has been published under the title *Form and Function*

(University of California Press, 1947). The definitive biography to date is by Nathalia Wright (*Horatio Greenough: The First American Sculptor.* Philadelphia: University of Pennsylvania Press, 1963).

17. H. T. Tuckerman, *A Memorial to Horatio Greenough* (New York, G. P. Putnam's Sons, 1853), p. 118.
18. Greenough, *Form and Function* p. 37. It is ironic that this is the same building which Robert Dale Owen, the director under whom it was commissioned and built, praised it as having liberated him from the "Procrustean" Classic!
19. *Ibid.*, p. 57.
20. *The Democratic Review*, vol. XIII, August, 1843.
21. Tuckerman, *Memorial*, p. 124.
22. Ralph Waldo Emerson, *English Traits* in *Works* (Boston: Houghton Mifflin, 1888), vol. V, p. 10. Here, as always, Greenough is urging the artist to use scientific knowledge as the basis for artistic decision.

CHAPTER 5

THE GOLDEN LEAP (pages 140–167)

1. *Magazine of Botany and Registry of Flowering Plants* (London, 1839), vol. IV, p. 11.
2. The definitive account of Paxton's professional career will be found in C. F. Chadwick's *Works of Joseph Paxton* (London: The Architectural Press, 1961). The best biography is that by his granddaughter Violet Markham, *Paxton and the Bachelor Duke* (London: Hodder, 1935).
3. *Magazine of Botany* (vol. VI, p. 12).
4. Markham, *op. cit.*, p. 182.
5. Chadwick, *Works.*
6. John Ruskin, *The Stones of Venice* (London: Library Edition, 1889), Appendix 17, p. 456. Elsewhere he called it "a magnified conservatory" and "a cucumber frame."
7. The definitive biography of Roebling, as well as the most thorough analysis of his work, is to be found in D. B. Steinman's *Builders of the Bridge* (New York: Harcourt, Brace and Company, 1945).
8. Roebling had clearly anticipated the phenomenon of aerodynamic instability in suspension bridges — the cause of the collapse of the ill-fated Tacoma Narrows Bridge in 1940 — in both his designs and his writings.
9. Gustave Eiffel, "The Eiffel Tower," *Report of the Smithsonian Institution* (Washington, D.C., 1889), pp. 729–35.
10. William A. Eddy, "The Eiffel Tower," *Atlantic Monthly Magazine* (June 1889), vol. LXIII, pp. 721–27.

CHAPTER 6

1860–1893: THE GREAT VICTORIANS (pages 168–213)

1. Boston: Houghton Mifflin, 1887.
2. *Looking Backward* (New York: Modern Library, 1942), p. 27.
3. *Ibid.*, p. 80.
4. *Architectural Record*, vol. I, no. 3 (January-March 1892), p. 268.
5. *Ibid.*, vol. IV, no. 1 (July-September 1894), p. 13.

6. Vincent J. Scully, Jr., *The Shingle Style: Architectural Theory and Design from Richardson to the Origins of Wright* (New Haven: Yale University Press, 1955).
7. Clarence Cook, "Architecture in America," *North American Review,* vol. CXXXV (Sept. 1882), pp. 243–252.
8. Thorstein Veblen, *The Theory of the Leisure Class* (New York: B. W. Huebsch, 1918), p. 154.
9. *Ibid.,* p. 349.
10. Henry-Russell Hitchcock, Jr., *The Architecture of H. H. Richardson* (New York: Museum of Modern Art, 1936).
11. Louis H. Sullivan, *The Autobiography of an Idea,* (Washington: American Institute of Architects, 1926), p. 200.
12. According to the excellent bibliography in *Louis H. Sullivan* by Hugh Morrison (New York: W. W. Norton, 1935), pp. 294–305.
13. See Carl W. Condit (*The Chicago School of Architecture.* University of Chicago Press: 1964) for an authoritative account of this subject.
14. Sullivan, *The Autobiography of an Idea,* p. 313.
15. Louis Henri Sullivan, *Kindergarten Chats* (New York: Scarab Fraternity Press, 1934), p. 194.
16. *Ibid.,* p. 8.
17. Sullivan, *Autobiography,* p. 314.
18. Frank Lloyd Wright, *Autobiography* (New York: Longmans, Green and Company, 1932), p. 124.
19. "Lessons of the Fair," *Cosmopolitan Magazine* (December 1893).
20. "A Farewell to the White City," *Cosmopolitan Magazine* (December 1893).
21. Montgomery Schuyler, *American Architecture and Other Writings,* edited by William H. Jordy and Ralph Coe (Cambridge: Harvard University Press, 1961), p. 557.
22. *Ibid.,* pp. 556–574.

CHAPTER 7

1893–1932: ECLIPSE (pages 214–253)

1. "A Bank Built for Farmers: Louis Sullivan designs a building which marks a new epoch in American architecture." *The Craftsman* (New York: November 1908) vol. 15, no. 2, pp. 183 ff.
2. Albert Bush-Brown, *Louis Sullivan* (New York: George Braziller, Inc., 1960), p. 28.
3. *Louis H. Sullivan, Democracy: A Man-Search* (Detroit: Wayne State University Press, 1961) pp. 272–73. Though written between 1906 and 1908, this series of polemical essays was not published until 1961. It was prepared for publication by Elaine Hedges who also wrote the illuminating Introduction.
4. Sullivan, *Autobiography,* p. 289.
5. Lewis Mumford, *Sticks and Stones* (New York: Dover Publications, 1955). New preface to the new edition, unpaged.
6. Grant Carpenter Manson, *Frank Lloyd Wright to 1910: The First Golden Age* (New York: Reinhold, Publishing Corporation, 1958), p. 154.
7. This material is still largely in periodical form. Cf. H. Allen Brooks, Jr. "The Early Work of the Prairie Architects," *Journal of the Society of Archi-*

tects, vol. xix (March 1960), no. 1, pp. 2–10; Mark L. Peish, *Walter Burley Griffin* (unpublished dissertation, Columbia University, 1959); David Gebhard, "Louis Sullivan and George Grant Elmslie," *Journal of the S.A.H.*, vol. xix no. 2, (May 1960), pp. 62–68; Esther McCoy, *Five California Architects* (New York: Reinhold Publishing Corporation, 1960).

8. *The Western Architect* (Minneapolis, 1915), vol. xxi, no. 1 (January); and vol. xxii, no. 1 (July).
9. See Esther McCoy, "Five Sullivan Letters," *Journal of the S.A.H.*, vol. xxii, no. 4.
10. For an excellent interpretation of the impact of the 1913 Armory Show on American art see Milton W. Brown, *The Story of the Armory Show* (New York: N.Y. Graphic Society, 1963).
11. William Pedersen, "Analysis of American Architectural Journals, 1919–1929" (unpublished paper; read at Modern Architecture Seminar, Columbia University, May 1962).
12. Henry-Russell Hitchcock and Philip Johnson, *The International Style* (New York: Museum of Modern Art. 193–).
13. William H. Jordy, "PSFS: Its Development and its Significance in Modern Architecture," *Journal of the Society of Architectural Historians,* vol. xxi, no. 2, May 1962, pp. 47–83.
14. *Ibid.*, p. 62.
15. *Ibid.*
16. *Ibid.*, p. 64.
17. Franklin Booth, *The Buildings of the Exposition: Their Architectural Meaning* (Winnetka: Book and Print Guild, 1934), unpaged.

CHAPTER 8

1933–1945: AMERICAN BUILDING AT THE CROSSROADS (pages 254–274)

1. Arthur Kingsley Porter, *Beyond Architecture* (Boston: Marshall Jones, 1918), p. 132.
2. Scammon Lectures, Chicago Institute of Art, 1915.
3. *New York Times Magazine* (April 21, 1940), p. 9.
4. Charles Harris Whitaker, *Rameses to Rockefeller* (New York: Viking Press, 1934).
5. For a more extended analysis, see Fitch, *Architecture and The Esthetics of Plenty* (New York: Columbia University Press, 1961), pp. 14–28.
6. Reinhard & Hofmeister; Corbett, Harrison & McMurray; Hood & Fouilhoux.
7. Quoted by Bernard Karpel, *Bibliographic Supplement to House Beautiful* (New York: November 1955), p. 17.
8. Fitch, *Architecture and The Esthetics of Plenty,* p. 110.
9. *Ibid.*, p. 141.
10. *Ibid.*, p. 156 ff.

CHAPTER 9

1945–1965: THE PARADOX OF ABUNDANCE (pages 275–317)

1. This symposium, as significant in its way as the Museum's famous exhibi-

tion *The International Style* of February 1932, was held on February 11, 1948.

2. *Bulletin,* Museum of Modern Art (New York, 1948), vol. XV, no. 3, p. 2.
3. *Ibid.,* p. 11.
4. S. Moholy-Nagy, *Architectural Forum* (New York), vol. 120, no. 2, p. 77.
5. James Marston Fitch, *Walter Gropius* (New York: Braziller, 1960), p. 13.
6. *Architectural Forum* (New York, May 1963), vol. 118, no. 5, p. 144.
7. See, for example, Le Corbusier's poignant autobiography *Creation Is a Patient Search* (New York: Praeger, 1960). See also my essay "Wright and The Fine Arts" *in Architecture and the Esthetics of Plenty* (New York: Columbia University Press, 1961).
8. Her book, *The Death and Life of Great American Cities* (New York: Random House: 1961), has proved to be one of the seminal books of the mid-century.
9. *Architecture and The Esthetics of Plenty,* pp. 216–226.
10. See Morton and Lucretia White, *The Intellectual and the City* (Cambridge, Mass.: Harvard University Press, 1964). See also Carl Schorske in *The City in History* (Cambridge, Mass.: Harvard University Press, 1964).
11. John D. Rosenberg, "The Case Against Citycide," *The New Leader* (New York, December 7, 1964), vol. xlvii, no. 25, pp. 9–11.
12. Edward Hudson, "When The Ceiling is Zero," *New York Times* (December 27, 1964), Sec. XX, p. 5.
13. Fitch, *Architecture and the Esthetics of Plenty,* pp. 221–223.
14. This section has been published, in somewhat altered form, in *The Professions* (Boston: American Academy of Arts and Sciences, 1965).
15. Letter, Joseph Watterson, editor, *Journal of the American Institute of Architects* (Washington), June 24, 1964.
16. Author's own estimate: at time of writing, AIA had not yet completed its first comprehensive survey of the profession.
17. *1963–1964 Report on Enrollment* (Washington: Association of Collegiate Schools of Architecture).
18. *List of Accredited Schools of Architecture: 1963–64* (Washington: National Architectural Accrediting Board, June 1, 1963).
19. Watterson, *Journal.*
20. Turpin C. Bannister, editor, *The Architect at Mid-Century: Evaluation and Achievement* (New York: Reinhold: 1954), Table 57, appendix.
21. *Ibid.,* Table 53.
22. John Galbraith, *The Affluent Society* (Boston: Houghton Mifflin, 1958), and Michael Harrington, *The Other America* (New York: Macmillan, 1962).
23. *Architectural Forum* (vol. 120, no. 4, April 1964, p. 38.
24. *Ibid.,* p. 97.
25. Henry Saylor, *The AIA's First Hundred Years* (Washington: The Octagon, 1957), p. 38.
26. Bannister, *The Architect at Mid-Century,* p. 4.
27. See my essay, The Forms of Plenty," *Columbia University Forum* (New York, Summer 1963), vol. VI, no. 3, pp. 4–9.

CREDITS FOR ILLUSTRATIONS

INDEX

NOTE: *Page references in italics refer to illustrations.*